# Advances in Industrial Control

**Other titles published in this series:**

*Digital Controller Implementation and Fragility*
Robert S.H. Istepanian and James F. Whidborne (Eds.)

*Optimisation of Industrial Processes at Supervisory Level*
Doris Sáez, Aldo Cipriano and Andrzej W. Ordys

*Robust Control of Diesel Ship Propulsion*
Nikolaos Xiros

*Hydraulic Servo-systems*
Mohieddine Jelali and Andreas Kroll

*Model-based Fault Diagnosis in Dynamic Systems Using Identification Techniques*
Silvio Simani, Cesare Fantuzzi and Ron J. Patton

*Strategies for Feedback Linearisation*
Freddy Garces, Victor M. Becerra, Chandrasekhar Kambhampati and Kevin Warwick

*Robust Autonomous Guidance*
Alberto Isidori, Lorenzo Marconi and Andrea Serrani

*Dynamic Modelling of Gas Turbines*
Gennady G. Kulikov and Haydn A. Thompson (Eds.)

*Control of Fuel Cell Power Systems*
Jay T. Pukrushpan, Anna G. Stefanopoulou and Huei Peng

*Fuzzy Logic, Identification and Predictive Control*
Jairo Espinosa, Joos Vandewalle and Vincent Wertz

*Optimal Real-time Control of Sewer Networks*
Magdalene Marinaki and Markos Papageorgiou

*Process Modelling for Control*
Benoît Codrons

*Computational Intelligence in Time Series Forecasting*
Ajoy K. Palit and Dobrivoje Popovic

*Modelling and Control of Mini-Flying Machines*
Pedro Castillo, Rogelio Lozano and Alejandro Dzul

*Ship Motion Control*
Tristan Perez

*Hard Disk Drive Servo Systems* (2nd Ed.)
Ben M. Chen, Tong H. Lee, Kemao Peng and Venkatakrishnan Venkataramanan

*Measurement, Control, and Communication Using IEEE 1588*
John C. Eidson

*Piezoelectric Transducers for Vibration Control and Damping*
S.O. Reza Moheimani and Andrew J. Fleming

*Manufacturing Systems Control Design*
Stjepan Bogdan, Frank L. Lewis, Zdenko Kovačić and José Mireles Jr.

*Windup in Control*
Peter Hippe

*Nonlinear $H_2/H_\infty$ Constrained Feedback Control*
Murad Abu-Khalaf, Jie Huang and Frank L. Lewis

*Practical Grey-box Process Identification*
Torsten Bohlin

*Control of Traffic Systems in Buildings*
Sandor Markon, Hajime Kita, Hiroshi Kise and Thomas Bartz-Beielstein

*Wind Turbine Control Systems*
Fernando D. Bianchi, Hernán De Battista and Ricardo J. Mantz

*Advanced Fuzzy Logic Technologies in Industrial Applications*
Ying Bai, Hanqi Zhuang and Dali Wang (Eds.)

*Practical PID Control*
Antonio Visioli

*(continued after Index)*

Lei Guo · Hong Wang

# Stochastic Distribution Control System Design

## A Convex Optimization Approach

Lei Guo, Prof.
Beihang University
Institute of Automation Science
and Electrical Engineering
Hai Dian District
37 Xueyuan Road
100083 Beijing
China
lguo@buaa.edu.cn

Hong Wang, Prof.
Northeastern University
Shenyang
China
*and*
University of Manchester
School of Electrical and Electronic
Engineering
Manchester, M60 1QD
United Kingdom
hong.wang@manchester.ac.uk

ISSN 1430-9491
ISBN 978-1-4471-2559-4
ISBN 978-1-84996-030-4(eBook)
DOI 10.1007/978-1-84996-030-4
Springer London Dordrecht Heidelberg New York

British Library Cataloguing in Publication Data
A catalogue record for this book is available from the British Library


Softcover reprint of the hardcover 1st edition 2010



*Cover design:* eStudioCalamar, Figueres/Berlin

Printed on acid-free paper

Springer is part of Springer Science+Business Media (www.springer.com)

# Advances in Industrial Control

Professor (Emeritus) O.P. Malik
Department of Electrical and Computer Engineering
University of Calgary
2500, University Drive, NW
Calgary, Alberta
T2N 1N4
Canada

Professor K.-F. Man
Electronic Engineering Department
City University of Hong Kong
Tat Chee Avenue
Kowloon
Hong Kong

Professor G. Olsson
Department of Industrial Electrical Engineering and Automation
Lund Institute of Technology
Box 118
S-221 00 Lund
Sweden

Professor A. Ray
Department of Mechanical Engineering
Pennsylvania State University
0329 Reber Building
University Park
PA 16802
USA

Professor D.E. Seborg
Chemical Engineering
3335 Engineering II
University of California Santa Barbara
Santa Barbara
CA 93106
USA

Doctor K.K. Tan
Department of Electrical and Computer Engineering
National University of Singapore
4 Engineering Drive 3
Singapore 117576

Professor I. Yamamoto
Department of Mechanical Systems and Environmental Engineering
The University of Kitakyushu
Faculty of Environmental Engineering
1-1, Hibikino,Wakamatsu-ku, Kitakyushu, Fukuoka, 808-0135
Japan

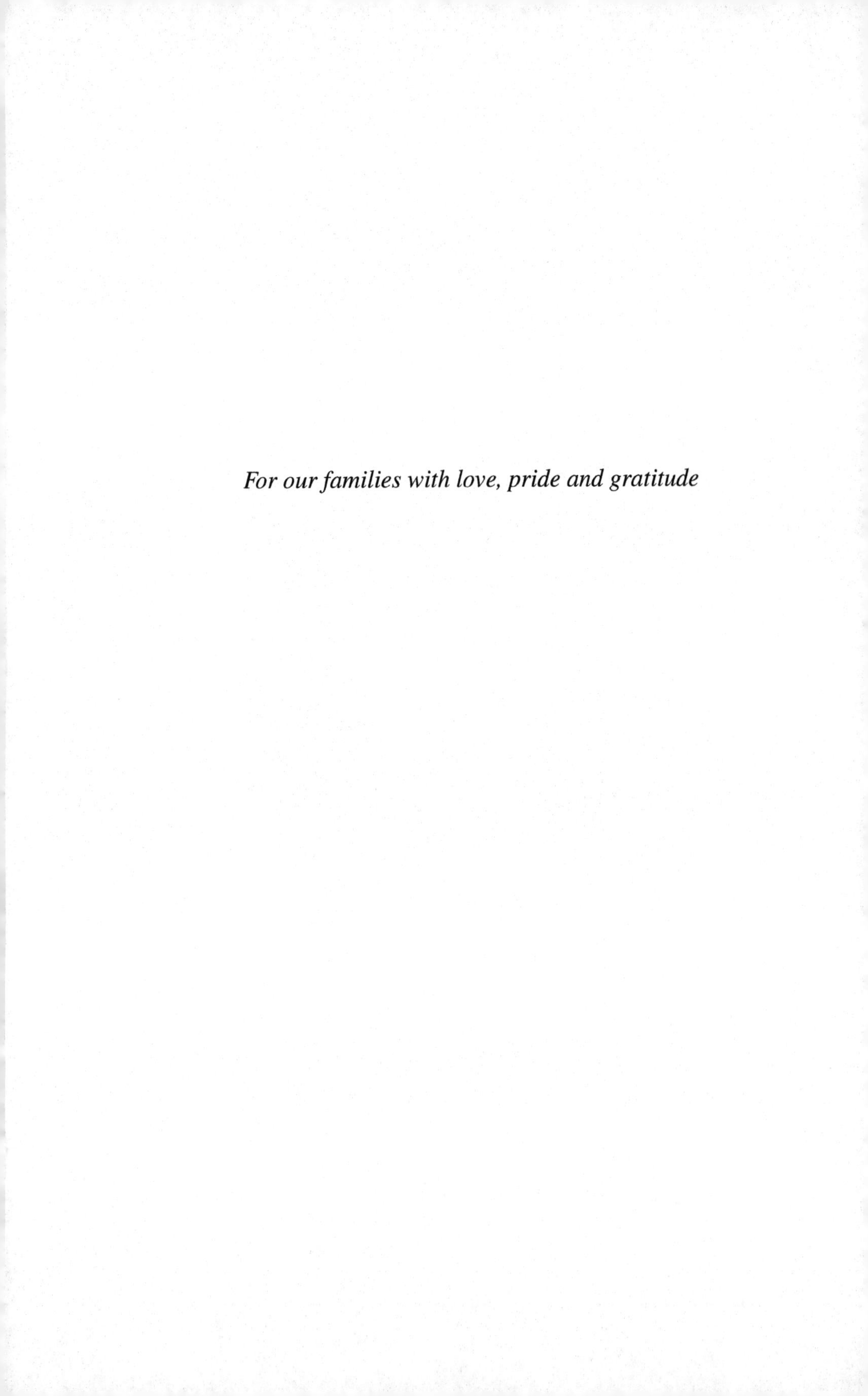

*For our families with love, pride and gratitude*

# Foreword

The series *Advances in Industrial Control* aims to report and encourage technology transfer in control engineering. The rapid development of control technology has an impact on all areas of the control discipline. New theory, new controllers, actuators, sensors, new industrial processes, computer methods, new applications, new philosophies, new challenges. Much of this development work resides in industrial reports, feasibility study papers and the reports of advanced collaborative projects. The series offers an opportunity for researchers to present an extended exposition of such new work in all aspects of industrial control for wider and rapid dissemination. Improving the performance and control of industrial processes can follow the three routes shown in Figure 0.1.

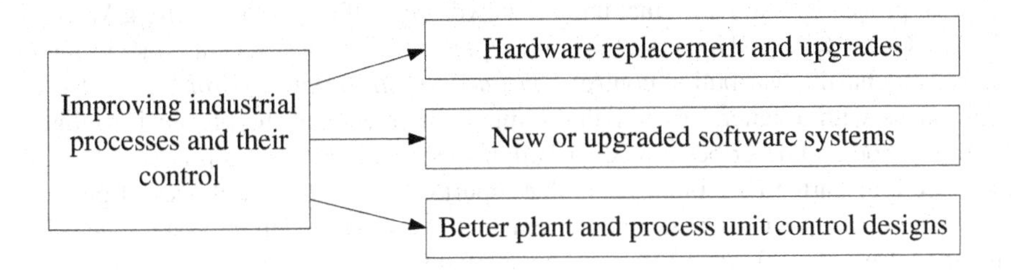


**Figure 0.1** Routes for enhancing the control of industrial processes

The installation of completely new process hardware, or the large-scale renovation of process hardware is usually capital intensive; consequently it is not undertaken frequently. Pro-active maintenance, however, is likely to be part of process plant operation and to be implemented continuously throughout the life cycle of the plant; the aim of this is to keep the process and its instrumentation working efficiently and effectively at all times to minimize or eliminate plant downtime.

New or upgraded software systems for plant operation are often less capital intensive and often produce results within a shorter payback period than hardware changes. Consequently, plants often undergo several software replacements or upgrades during their lifetime. New software systems often include revised supervisory control systems and new low-level control modules. Revised and enhanced operator interfaces and displays help operators to improve control and understand process behavior better. Add-ons may include process-monitoring modules and online expert systems to support operator decision making.

Perhaps the least expensive route to enhanced process performance is to consider how to obtain the best control with the existing process hardware, software, and instrumentation configuration. Such a review could look at the process control hierarchy that exists on the plant and optimize control performance at each level. At the lower levels of control, it is highly likely that the controllers have been tuned to meet classical time domain and frequency domain specifications. Alternatively, other controller specifications based on a set of statistics like the output mean and the output variance may have been used with the assumption that the underlying stochastic processes are Gaussian. Control design philosophies like minimum variance and linear-quadratic-Gaussian control aim to capture these stochastic viewpoints. However, for some processes other controller specification frameworks suit the process characteristics better. This idea was taken up by Hong Wang (University of Manchester, UK) who was faced with the problem of designing controllers for paper making processes where it is desired to have the 2D density distribution of paper fibers as close as possible to a uniform distribution. This practical problem provided the impetus to develop a new group of control design techniques based on non-Gaussian assumptions.

From then on, stochastic distribution control turned out to be a fruitful and systematical field of control system research. The idea has particular relevance to the process industries where non-Gaussian output probability density functions often occur in processes like paper-making and powder manufacture. Now, Hong Wang and Lei Guo (Beihang University, China) report on their most recent work in the field of stochastic distribution control. *Stochastic Distribution Control System Design*, opens with a general review that reminds the reader of the physical origins of the method and describes how distribution specifications arise in practice. Then there are four parts to the monograph that report on four different classes of problems. These include innovative and systematical work on modeling, system analysis, fixed structure (PID) controller design and fault detection and diagnosis. In each chapter, illustrative examples have been given to demonstrate the outcomes of the various design procedures proposed.

This monograph demonstrates the theoretical depth that has been achieved in the intervening years by the authors. The book will appeal to academic researchers, postgraduate students and industrial engineers seeking more effective controller designs for processes with non-Gaussian characteristics.

Industrial Control Centre,
Glasgow
Scotland, UK
2009

*M.J. Grimble*
*M.A. Johnson*

# Preface

Stochastic distribution control aims to control the shape of the output probability density functions (PDFs) for non-Gaussian and dynamic stochastic systems. This differs from traditional stochastic control where only the output mean and variance are considered. Stochastic distribution control was originally developed by the second author of this book when he considered a number of challenging paper making modeling and control design problems in 1996. For example, the solid distribution control of a 3D paper web is required during the initial paper web forming phase using a number of controllable process variables in the wet end systems. Since 1996, many modeling and control algorithms have been developed with application to material processing, mineral processing, combustion control and paper making systems. For example, a closed loop real-time paper web formation control has been established for the first time in paper making.

In 2000, the monograph named *Bounded Dynamic Stochastic Systems*, written by the second author of this book, was published by Springer-Verlag London, where a forwarded message from Professor Michael Grimble and Professor Mike Johnson stated that stochastic distribution control is "*a far more stringent control specification than say output mean and variance measures alone.*". Indeed, this research has been regarded as a new branch in stochastic control. At present there are more than 20 research centers world-wide working in this area. Invited sessions have been seen at major international conferences (e.g., the 2002 American Control Conference and the 2004 IEEE Symposium on Intelligent Systems and Intelligent Control) together with several special issues published by refereed international journals.

More recently, it has been observed that general data-driven dynamic modeling, data dimensional reduction and filtering algorithms design can all be transformed into a PDF shaping and control design problem. For example, in modeling unknown systems, the best model can be obtained by selecting model parameters (i.e., neural network weights) so that the PDF of the modeling error follows a narrowly distributed Gaussian with zero mean. In filtering algorithms design, the best filtering results can be achieved by selecting filtering parameters (i.e., filter gain matrix) so that the filtered residual's PDF can again be made to follow a narrowly distributed

Gaussian with zero mean. It can therefore be expected that many applications of the stochastic distribution control will occur in the near future.

With the developments in instrumentation, computer science, data and image processing technologies in many practical processes, one can easily obtain the full set of statistics (moments and entropy) or the PDF from either the data set or images of the measured output. Another new feature of the recent stochastic distribution control methodologies is that not only is the control objective concerned with the PDF, but the driven information for feedback control is also characterized by PDF or the statistical information set. This means that stochastic distribution control has developed a new branch as a type of data-driven stochastic control, which theoretically differs from most of the previous control approaches.

Following the funding support from Chinese National Science Foundation (Grants No 60935012, 60828007, 60534010, 60774013, 60474050), the Chinese 863 project, the Chinese 973 project (2009CB320601, 2009CB320604), the Chinese 111 project (B08015) and the UK Leverhulme Trust (F/00038/D and F/00120/BC), UK EPSRC (EP/G059837), since 2002, a lot of work has been carried out by the authors using linear-matrix-inequalities (LMI) based convex optimization approaches to solve various control design problems encountered in stochastic distributions systems. A number of technical papers have been published in leading control journals and conferences. In fact, a fruitful and systematical stochastic distribution control, filtering and fault detection framework has recently been established. In this framework, with the established new stochastic distribution system models, LMI-based convex optimization has been effectively used as the system analysis and synthesis tools. On the other hand, the applications to stochastic distribution control also lead to many challenging problems about LMI-based control and filtering algorithms. In this context, we feel that it would be beneficial to write a book on the particular subject of using LMIs to design stochastic distribution systems.

The book begins with a summary of the state of the art developments in stochastic distribution control, and then focuses on the detailed descriptions of various LMI-based configurations used for output PDF control, filtering and fault detection for non-Gaussian systems. We therefore hope that readers will obtain a full picture of the latest developments in stochastic distribution control fields, and that new LMI-based design techniques for complex systems with uncertainties, nonlinearities, constraints and time delays will be described.

The authors would like to thank their families and friends for their consistent support during the writing of the book. In particular, the first author would like to thank his parents, his wife and his son (Xiaoyu Guo). He would also like to thank his PhD student Yang Yi, who has made great contributions to the editing of the book. The second author would like to thank his son (Michael Wang) and daughter (Meijie Wang) for their understanding and support.

Beihang University, Beijing, PR China
Northeastern University, Shenyang, PR China
and the University of Manchester,
Manchester, UK
April, 2009

*Lei Guo*
*Hong Wang*

# Contents

# Abbreviations and Notation

Throughout this book, the following conventions and notation are adopted:

| | |
|---|---|
| SDC | Stochastic distribution control |
| NN | Neural network |
| LMI | Linear matrix inequality |
| PDF | Probability density function |
| PDE | Partial differential equation |
| PID | Proportional integral derivative |
| SISO | Single input and single output |
| MIMO | Multiple input and multiple output |
| DNN | Dynamic neural network |
| STC | Statistic tracking control |
| SIS | Statistic information set |
| FDD | Fault detection and diagnosis |
| VSC | Variable structure control |
| LQG | Linear-quadratic-Gaussian |
| FTC | Fault tolerant control |
| DCS | Distributed control systems |

| | |
|---|---|
| $R$ | The field of real numbers |
| $R^n$ | The set of $n$-dimensional real vectors or tuples |
| $R^{n\times m}$ | The set of $n \times m$-dimensional real matrices |
| $I(I_n)$ | The identity matrix (of dimension $n \times n$) |
| $x^T$ or $A^T$ | The transpose of vector $x$ or matrix $A$ |
| $A^{-1}$ | The inverse of matrix $A$ |

| | |
|---|---|
| $P > 0$ | Matrix $P$ is real symmetric positive definite |
| $P \geq 0$ | Matrix $P$ is real symmetric semi-positive definite |
| $P < 0$ | Matrix $P$ is real symmetric negative definite |
| $P \leq 0$ | Matrix $P$ is real symmetric semi-negative definite |
| $\lambda(A)$ | The set of eigenvalues of $A$ |
| $\lambda_{max}(A)$ | The maximum eigenvalues of real symmetric $A$ |
| $\lambda_{min}(A)$ | The minimum eigenvalues of real symmetric $A$ |
| $\|\|x\|\|$ | The Euclidean norm of vector $x$ |
| $\|\|A\|\|$ | The induced Euclidean norm of matrix $A$ |
| $\|\|x\|\|_p$ | The $l_p$ norm of vector $x$ |
| $\|\|A\|\|_p$ | The induced $l_p$ norm of matrix $A$ |
| max $S$ | The maximum element of set $S$ |
| min $S$ | The minimum element of set $S$ |
| sup $S$ | The smallest number of that is larger than or equal to each element of set S |
| inf $S$ | The largest number of that is smaller than or equal to each element of set S |
| sgn | The signum function |
| $[a,b)$ | The real number set $\{t \in R : a \leq t < b\}$ or the integer set $\{t \in N : a \leq t < b\}$ |
| diag$(A)$ | The diagonal matrix of matrix $A$ |
| sym$(A)$ | The value of matrix $A + A^T$ |
| tr$(A)$ | The trace of matrix $A$ |

# Chapter 1
# Developments in Stochastic Distribution Control Systems

## 1.1 Introduction

Stochastic systems are widely encountered in control engineering design. This is because almost all control systems are subject to random signals such as those originating from system parameter variations and sensor noise, etc. As such, one of the important practical issues in controller design is to minimize the randomness in the closed-loop system. This has motivated studies on minimum variance control [3] whose purpose is to minimize variations in the controlled system outputs or the tracking errors. Indeed, even today, most of the stochastic control design has focused on the control of the output mean and the variance of stochastic systems. In general, these developments are largely based on the assumptions that the system variables are of Gaussian types. Such assumptions, although strict, allow control engineers to make use of well-established stochastic theory to perform controller design and closed-loop system analysis. Indeed, the rather rich literature about the treatment of Gaussian systems has always been regarded as a basis for the design and analysis of stochastic control systems.

A random variable $x$ is said to obey a Gaussian distribution if its probability density function (PDF) is described as follows

$$\gamma(x) = \frac{1}{\sqrt{2\pi}\sigma} \mathrm{e}^{-\frac{(x-\mu)^2}{\sigma^2}}$$

where $-\infty < x < \infty$ is the range of variable $x$, $\mu$ and $\sigma^2$ are the mean value and variance, respectively. Such a PDF is shown in Figure 1.1, where it can be seen that this PDF is purely defined by the two parameters, $\mu$ and $\sigma^2$.

In industrial processes such as those for steel and paper making and general material processing, product quality data can be approximated by the above Gaussian PDF when the system operates normally. For example, in paper making, one product quality parameter is the grammage of finished sheets, which obeys a Gaussian distribution if the machine operates normally. Also, the strength of the finished paper can be characterized by a Gaussian PDF under nominal operating conditions. However,

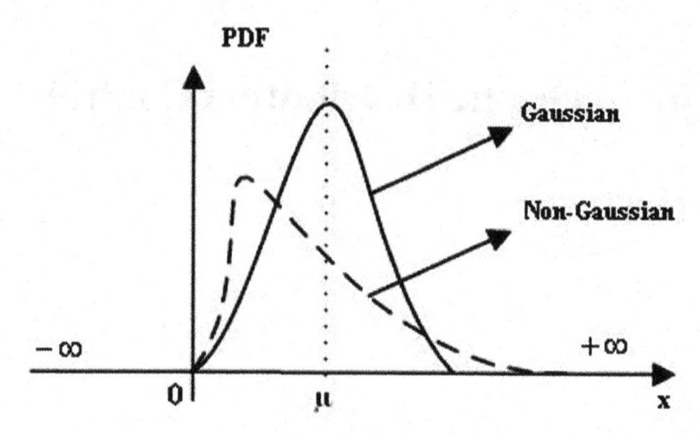


**Figure 1.1** The Gaussian PDF $\gamma(x)$

when abnormality occurs along the production line, these quality variables will not be Gaussian. In this regard, actions need to be taken so that the control variables along the production line can be tuned to bring these quality variables back to Gaussian distributions. Strictly speaking, hardly any variables in control systems obey a Gaussian distribution. This is simply because most random variables and random processes are bounded. As such, only approximations to a Gaussian distribution can be made.

Such approximations became the mathematical fundamental pre-request and assumption for the development of stochastic control in the past, where Ito's differential equation of the following form was used to characterize the dynamic:

$$dx = f(x,u)dt + g(x,u)d\omega$$

where $x$ is a random process, $u$ is a control input, $f(x,u)$ and $g(x,u)$ are the system dynamics, respectively, and $\omega$ is the well-known Brownian motion. With such a wonderful Brownian motion, it has been shown that $d\omega$ is a Gaussian random process. Indeed, it is the above Ito's differential equation that forms the mathematical basis for stochastic differential equation theory and also stochastic control. However, since in practice hardly any variables strictly obey a Gaussian distribution, the study of non-Gaussian dynamic systems is imperative.

PDF shape control has been studied for non-Gaussian and dynamical stochastic systems in response to the increased requirements from many practical production systems [11, 15, 22, 24, 25, 28, 30–32, 34, 36–39, 42, 44–46, 58, 61–69, 72, 73, 78, 80–83, 87–89, 92, 93, 97–99, 102, 108, 109, 112, 119, 120, 125, 128, 129, 132, 134–136, 138, 141–143, 145, 146, 155–169, 172, 173, 178, 181–188] where, for this type of system, the actual controlled output is the shape of its output PDFs and the inputs are only related to time. The system is shown in Figure 1.2, where the controller design problem formulates the control input $u(t)$ such that the output PDF is made as close as possible to a required distribution shape. Such a required distribution shape is represented by a PDF, and is refereed to as either the target distribution or target PDF throughout this book (see Figure 1.5). In practice, such a target PDF can be obtained in the same way as the determination of the set-points used for general closed-loop control. For example, for molecular weight distribution (MWD) control systems in chemical engineering, the target distribution is determined by the actual strength requirement of the products. For plastics a two-peak PDF shape is always used.

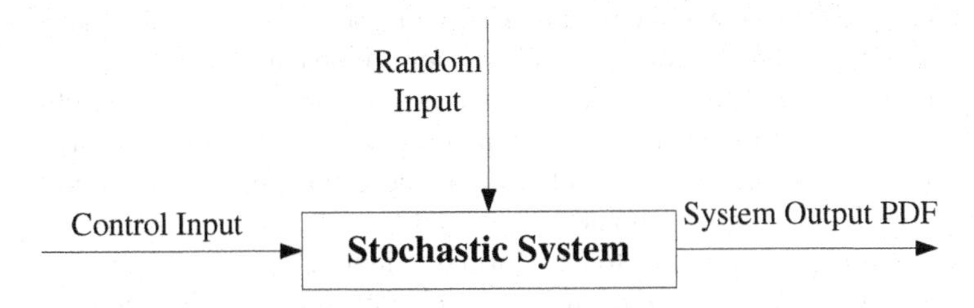


**Figure 1.2** The structure of a stochastic distribution control system

In terms of system modeling, this means that the controlled output PDF is a function of both time and space and can therefore be expressed as the following partial differential equation (PDE):

$$0 = \Xi\left(\frac{\partial^n \gamma}{\partial y^n}, \frac{\partial^{n-1}\gamma}{\partial y^{n-1}}, \ldots \frac{\partial \gamma}{\partial y}, \gamma, \frac{\partial^m{}_t}{\partial_t^m}, \frac{\partial^{m-1}{}_t}{\partial_t^{m-1}}, \ldots \frac{\partial_t}{\partial_t}\right) \tag{1.1}$$

where $\Xi(\ldots)$ is a general nonlinear function and $\gamma(.)$ denotes the output PDF. In theory, this PDE model can be established from the general expression of many population balance equations such as the widely used particulate system model [11, 15, 22, 28, 30, 32, 34, 36, 42, 44–46, 78, 80–83, 89, 93, 97–99, 102, 108, 112, 119, 120, 125, 128, 129, 132, 135, 138, 142, 143, 145, 178, 185], where the control design objective is to ensure that the output PDF's shape follows a target distribution shape. This objective is in high demand in many industrial systems, such as chemical engineering [3, 9, 12, 14, 23, 26, 27, 29, 33, 52, 81, 90, 95, 100, 117, 139, 140, 153, 154], powder manufacture [12], paper making [139], and combustion flame distribution processes [146], etc. Therefore, the general PDE model should be used but, in practice, it is generally difficult to obtain such a PDE model. Even if such a PDE is available, control design is still difficult to perform. As such, a new group of techniques have been developed for the modeling and control of output PDFs for general stochastic systems.

### *1.1.1 Paper Web Formation Systems*

The stochastic distribution control (SDC) problem originated when a number of systems in pulp and paper making were considered. For example, in the wet end of paper machines, there are normally several sections in which the initial paper web is formed. These sections are referred to as the head-box approaching system, the head-box and the moving wire table. In the head-box approaching system and the head-box, fiber, fillers and other chemical additives are mixed. This mixture generally consists of 5% solids and 95% water. When this mixture is injected onto the moving wire table, some water is drained through the wire into a white water pit underneath the wire table, leaving a fiber-dominant network on the wire table, which is regarded as the paper web. Due to the random nature of fiber length and filler particles, the solid density distribution of the web is random. When an image analysis based sensor is used, such a density distribution can be measured and represented as a 2D grey level distribution [139, 156, 161], which is generally controlled by the thick stock input and some chemical inputs before the head-box. To achieve good web formation, these control inputs should be selected to make the fiber network density distribution as close as possible to a uniform distribution.

In an early study of formation control using the idea of SDC, the distribution of optical basis weight was taken as the system output to be controlled. The input variables used were the addition of two flocculating agents [87]. The aim of control was to keep the same mean mass density of the paper sheet and make it more homogeneous. In 2002, the implementation of an on-line closed-loop formation control system for the pilot machine at University of Manchester Institute of Science and Technology (UMIST) was made. Recently such a web formation control strategy has been implemented at the paper board machine in Iggesund (see Figure 1.3), in which the gray-level distribution of sheet image is used to represent the mass density distribution [186]. A digital camera captures images of the moving sheet and the gray-level distribution as shown in Figure 1.4 can be abstracted from the image by certain image analysis techniques so as to form a PDF as shown in Figure 1.5. Different statistical measures such as variance, entropy and PDF shape have been evaluated by on-line closed-loop operations.

### *1.1.2 Flame Distribution Control*

The combustion system shown in Figure 1.6 serves as another example. The system is subjected to a fuel input together with some operating conditions, and produces a flame (or a temperature) distribution inside the combustion chamber. With the developments in image processing, several digital cameras can now be used to measure the distribution of the flame, which can be further transformed into the temperature distribution [146]. An efficient combustion means that the temperature distribution shape needs to be controlled. Since the distribution of flames is directly related to the distribution of temperature, this can also be stated as controlling the fuel flow rate so that the flames distribution (which can be represented as a multi-dimensional PDF) can be made to follow a target distribution. The key advantage of using flames distribution control in practical situations is that a fast closed-loop control can be established. This overcomes the difficulties caused by long time delays for exist-

**Figure 1.3** A paper board machine

**Figure 1.4** The 2D gray-level image

ing combustion control systems seen in power generation and other applications. Another advantage of using SDC here is that the combustion process can be made more effective than the existing ones, leading to significant fuel savings (such as coal powders, etc.).

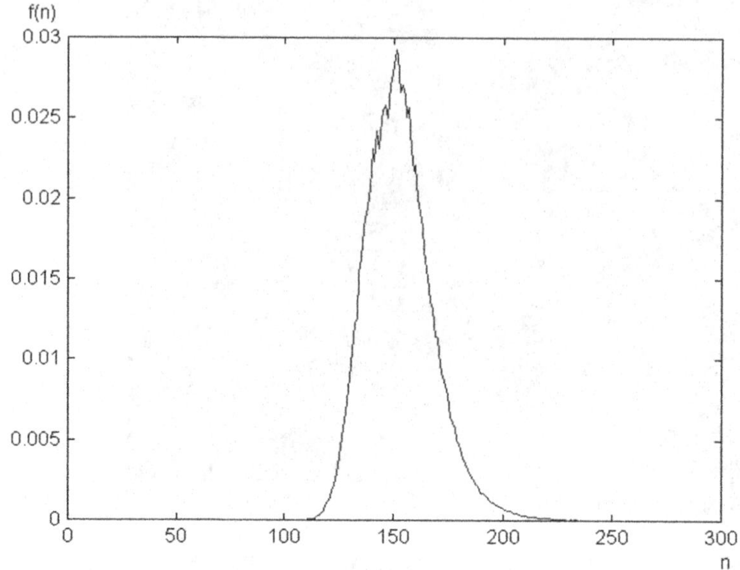


**Figure 1.5** The PDF shape of the image in Figure 1.4

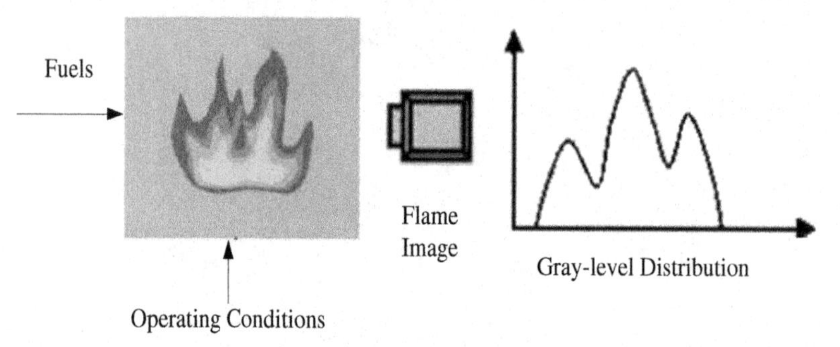


**Figure 1.6** Flame distribution control system

## *1.1.3 Challenging Issues*

The above two systems, as well as many other stochastic PDF shaping control systems, can generally be represented as in Figure 1.2, where $u$ is the control input to the system and $\gamma(y,u)$ is the output PDF of the system. The system is also subjected to random inputs that come from physical effects, such as raw material properties. To effectively control the output PDF shape, it is important to investigate the modeling and control issues of such systems. In terms of modeling, it is necessary that the model obtained should be easily accommodated in an on-line control framework. In this context, general PDE models are inappropriate. Once the model has been developed, controller design can be performed that produces an optimal control $u$ so that the output PDF is made to follow its target PDF as closely as possible.

This type of control is called SDC, which is a new research area where simple and implementable solutions for industrial processes are needed. In comparison with the traditional stochastic control theory, where only output mean and variances are of concern, SDC can offer an improved solution and is not restricted to just Gaussian input cases.

With the improvements in sensing technology, the output PDFs of these systems are now becoming measurable [24, 25, 31, 37–39, 58, 61–69, 72, 73, 87, 88, 92, 109, 134, 136, 141, 146, 155–169, 172, 173, 181–184, 186–188, 196]. This has provided an excellent opportunity for control engineers to develop, for the first time, direct feedback control of the output PDFs, leading to a much improved closed-loop system performance. For example, a 2D distribution of a paper sheet [188] is shown in Figure 1.4. This mimics a visual observation of someone holding a piece of paper against a strong light source. A good paper sheet would have a 2D uniform solids distribution. Therefore, this is an output PDF control problem where the inputs are the chemical and mechanical variables in the paper machine while the output is the PDF of the gray-level distribution. In this regard, the target distribution should be represented as a narrowly distributed Gaussian shape.

It can therefore be seen that the problem of controlling the output PDF has been investigated extensively by researchers in the control community in recent times. The first paper was published in *Automatica* in 1996 by Karney [92], where the process was represented by a PDF and the control is also expressed in a PDF form. As such, the purpose of controller design is to obtain the PDF of the controller so that the closed-loop PDF will follow its target PDF. However, as the control input is only related to time, the realization of the controller PDF was challenging. Therefore, the first practically implementable control strategy was developed by Wang at UMIST in 1996 [157]. Since then, rapid developments have been seen and at present there are around 20 research groups in the world actively seeking solutions to the controller design and their applications. Special sessions and special issues have been seen in various international control conferences and refereed journals. In 2000, a book by the second author of this monograph was published by Springer-Verlag, London [157], summarizing the earlier developments in this new research area. From then on, SDC has been classified into the following four approaches:

1. output PDF control using NN;
2. output PDF control using system input-output models;
3. minimum entropy control for non-Gaussian stochastic systems;
4. applications of the SDC concept to filtering design and fault diagnosis.

In the following sections, these approaches will be reviewed.

## 1.2 Stochastic Distribution Control when Output PDFs are Measurable

In terms of methodologies, the SDC control design algorithms can generally be classified into two groups, namely when the output PDFs are measurable and when the output PDFs are not measurable. This section will review a group of control algorithms for measurable output PDF systems.

In fact, the use of NNs for output PDF control is included in the first group of approaches for SDC design, where the NNs are used to approximate the instantaneous and measurable output PDFs and the NN parameters (i.e., weights and biases) are dynamically linked to the control input [58, 61, 63, 64, 146, 157, 160, 164, 166, 168, 173, 181–184]. In this regard, the following simple B-spline approximation was originally used in SDC to approximate the system output PDF:

$$\gamma(y,u_k) = \sum_{i=1}^{n} w_i(u(k))B_i(y) + e(y,u_k); \forall y \in [a,b] \tag{1.2}$$

where $u_k$ is the input at sample time $k$, $w_i(u_k)$ are the weights of the approximation to the output PDF, $\gamma(y,u_k)$, and $B_i(y)$ are the fixed basis functions. However, since the integration of $\gamma(y,u_k)$ over its definition domain $[a,b]$ is always 1, only $n-1$ weights are independent. Letting $V_k$ denote the vector of independent weights, then the dynamic part can be expressed as

$$V_{k+1} = f(V_k,u_k) \tag{1.3}$$

where $f(V_k,u_k)$ is a vector function that represents the dynamics between the NN weights vector and the control input. As a result, Equations 1.2 and 1.3 constitute the general modeling structure for stochastic distribution systems, where the input is a time-varying signal and the output is an output PDF [157]. This model differs from existing models (i.e., Ito's differential equations) used in stochastic control, which are only valid for Gaussian input. For the system represented by Equations 1.2 and 1.3, the purpose of controller design is to realize output PDF shape control. For multiple input and multiple output (MIMO) systems, the output PDFs can be a combined PDF of all the output variables.

Once the structure of the NN is selected, the control of the PDF shape can be regarded as the control of the weights and biases. For model (1.2) and (1.3), this output PDFs control can be realized via the control of $V_k$. As such, many existing control methods can be directly used to formulate the required weight control laws as represented by Equations 1.2 and 1.3. In particular, when the dynamics are linear (i.e., $f(V_k,u_k) = AV_k + Bu_k$) and the output PDF is approximated by a B-spline NN (Equation 1.2), a relatively simple solution has been established which minimizes the following performance index [157, 164]:

$$J = \sum_k \int_{\Omega\in[a,b]} (\gamma(y,u_k) - g(y))^2 \mathrm{d}y + u_k^T S u_k \tag{1.4}$$

where $g(y)$ is the target PDF, $u_k$ is the control input and $R$ is a weighting matrix that constrains the input. The integration is calculated over $\Omega$ on which the output PDF is defined, and the first term represents the cumulative functional distance between $\gamma(y,u_k)$ and $g(y)$. Based on Equations 1.2-1.4, a number of control algorithms have been developed to control the output PDF shape. A linear feedback control was originally established in [157], where a feedback control law was developed that used a linear combination of the measured PDFs and past inputs as feedback for closed-loop control. This was then followed by the development of control algorithms for nonlinear systems [160]. Moreover, analysis of the closed-loop robustness has also been performed in [61, 63, 157, 164]. When the weight dynamics $f(V_k,u_k)$ are unknown and linear, a new scanning, recursive parameter estimation algorithm has been developed that provides an estimate of linear parameters in model (1.3) (see [157]).

Depending on the dimensions of the system, the NNs that are used to approximate the output PDFs can be either multi-layer perceptions (MLPs) or radial basis functions (RBFs). Furthermore, to improve the accuracy of the SDC models and to enhance the numerical robustness of the NN approximations to the output PDFs, several B-spline NN methods have been developed and examples are the square root B-splines PDF model and rational B-splines PDF model [64, 166, 167]. The advantage of using square root B-spline approximation is that the calculated output PDFs can always be guaranteed to be positive. On the other hand the use of the rational B-spline approximation can enlarge the operational domain of the approximation to the output PDFs.

Originally, controllers were designed by using numerical solutions that minimize Equation 1.4. This led to control algorithms with a high computational load for closed-loop realization. Moreover, there are other theoretical and practical obstacles in the application of the previous results:

1. Stability analysis is hard to perform so that the closed-loop systems cannot be guaranteed to be stable.
2. Only linear models are considered, which usually cannot change the shape of input PDFs.
3. Constraints of the weight vectors are ignored so that the tracking diverges.

To overcome these problems, fixed control structures, such as PID, have also been developed [61, 63, 64, 183, 184], where linear matrix inequalities (LMIs) are used to obtain the required control parameters. The advantage of using a fixed structure controller is that the total number of control parameters involved can be minimized and then designed off-line, leading to simplified control algorithms design with guaranteed closed-loop stability. Using LMI techniques, an integrated analysis and synthesis framework can be established.

The key requirement for this group of methods is that the output PDF is measurable. This requirement is met in many practical processes where the PDF can be obtained from either measured data or images. There is also a new feature which differs from any other previous feedback control strategies with measured and determined values of the output. This generates a feedback control law where the weighted and measured output PDFs are used as the feedback control signals.

Once the closed-loop system is formulated, stability and performance analysis can be readily carried out using standard stability analysis tools such as Lyapunov theory (see [58, 63, 64, 88, 164, 183]).

SDC systems have their output represented as output PDFs while the inputs are normal time dependent variables. Therefore, the closed-loop system has the general structure shown in Figure 1.7, where a typical response for the closed-loop system is represented by the 3D plot shown in Figure 1.8.

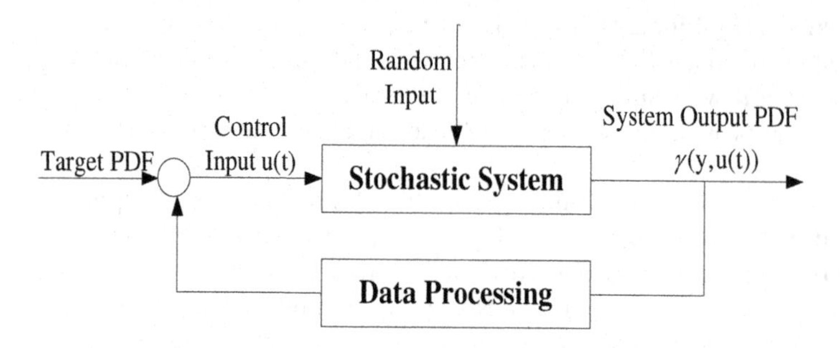


**Figure 1.7** The structure of a closed-loop system

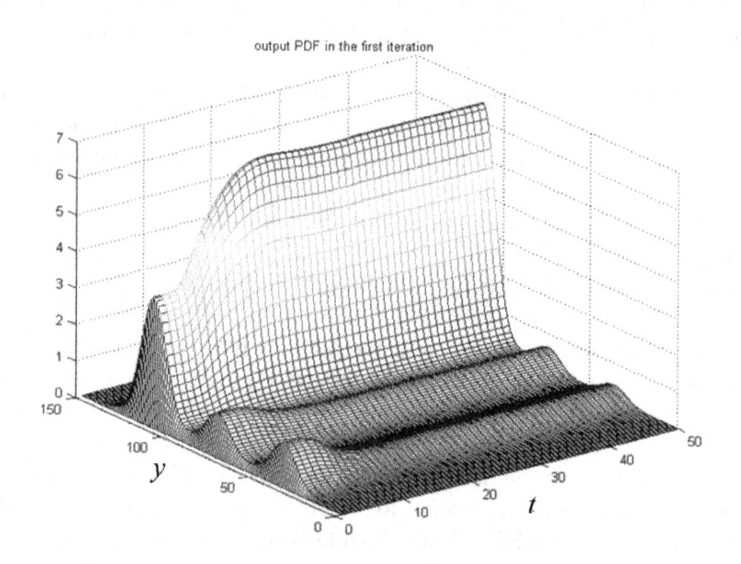


**Figure 1.8** A typical closed-loop response for SDC systems

As shown in Figure 1.8, initially the output PDF has one dominant peak. After closed-loop control for a period of time from 0-50 seconds, the output PDF shape changes and has three identifiable peaks.

## 1.3 Stochastic Distribution Control when Output PDFs are Unmeasurable

The main problem with the NN approach described in the previous section is that the model used for control design does not have a direct physical meaning. Also, the size of the NN can be very large if the output PDF shape is complicated in terms of having a large number of peaks. This leads to high dimensional dynamics between the NN weights and the control input (see model (1.3)). To solve this problem, a general input and output physical model of system has been used:

$$y_k = f(y_{k-1}, y_{k-2}, \cdots, y_{k-n}, u_{k-d}, u_{k-d-1}, \cdots, u_{k-m}, w_k) \tag{1.5}$$

where $y_k$ is the system output, $u_k$ is the input, $w_k$ is a random process whose PDF (denoted by $\gamma_w(x)$) is assumed known and $f(...)$ characterizes the dynamics of the system. The output PDF can then be formulated as a function of the PDF of the random input and all the past input and output measurements. As a result, the control input can be readily obtained using standard optimization routines for the performance function [158, 187]. In this context, the control input design is carried out in two logical steps:

1. Formulate the PDFs of the output in Equation 1.5 using function $f(...)$ and the PDF of the noise term.
2. Minimize the performance function in Equation 1.4.

The first step involves the use of probability theory where the standard formulation of the PDF of a function of a random variable is required. This can be broken down into two cases. In the first case it is assumed that function $f(...)$ is invertible with respect to its noise term $w_k$. The PDF of the system output $y_k$ is then given by:

$$\gamma(y, u_k) = \gamma_w(f^{-1}(\phi_k, u_k, y)) \left| \frac{df^{-1}(\phi_k, u_k, y)}{dy} \right| \tag{1.6}$$

where $\phi_k = [y_{k-1}, y_{k-2}, \cdots, u_{k-1}, u_{k-2}, \cdots]^T$ groups all the past inputs and outputs of the system represented by Equation 1.5. Using Equation 1.6, control algorithms can be readily developed by minimizing the performance index (1.4), which leads to a numerical solution for the required closed-loop control.

This was summarized in [158], where a recursive formula for $\gamma(y, u_k)$ was established that links directly to the control input. This was then followed by the development of the required control algorithm to minimizes Equation 1.4. A closed-loop stability analysis was also performed so that stable system operation could be guaranteed. However, since the recursive formula for the output PDF is based on a first order approximation, the accuracy of the closed-loop control needs to be carefully considered. This led to further developments for general MIMO systems using Equation 1.5 as reported in [69], where a predictive form of the output PDF was formulated. This was followed by the introduction of the new concepts of hybrid probabilities and hybrid random variables. Using these new concepts, a simple controller design has been obtained that realizes the desired tracking of the output PDF and ensures the stability of the closed-loop systems. To summarize, the advantage of using Equation 1.6 to formulate the output PDF control is that there is no need to measure the output PDF so long as the model in Equation 1.5 is accurate. This means that a good robust solution will be necessary if the model is not accurate. As such, further developments are required to obtain an effective output PDF control algorithm.

To address the problem of (unmeasurable) output PDF control PDF for stochastic systems with random parameters, a new identification algorithm has been developed that uses input and output measures of stochastic systems to estimate on-line the unknown PDF of system parameters. In this context, input and output models have again been used as a starting point, where generating functions from probability theory have been used to transform the output PDF into a simple algebraic form. This was then followed by the development of a scanning least squares algorithm [165]. Once these PDFs for the unknown parameters were estimated, output PDF control was developed [173].

In this book, the data-driven SDC and statistic tracking control approaches have been presented which generalized the previous works (see [68, 181, 183]).

## 1.4 Minimum Entropy Control

In the above sections, the control algorithms have been designed under the assumption that the target PDF is available. When the target PDF is not available, the aim of closed-loop control for the stochastic systems should be to reduce the randomness of the system. Since entropy is a general measure of the degree of randomness/uncertainty of random variables, entropy can be used as the performance index to be minimized during control design. This has led to recent developments in minimum entropy control for SDCs [155, 162, 186, 187, 196] and the performance to be minimized is defined as follows

$$J = -\sum_{k} \int_{\Omega} \gamma(y, u_k) \log[\gamma(y, u_k)] \mathrm{d}y + u_k^T R u_k \tag{1.7}$$

where the first term is the output PDF entropy, and the other terms are the same as those in Equation 1.4.

In minimum entropy control, the control algorithms can be classified into two cases, one for measurable output PDFs (see [155, 162, 186]) and the other for unmeasurable PDFs (see [188]). When the output PDF is measurable, the NN approaches can still be used where the NN approximate the measured output PDF and the dynamics of the system are represented by a set of differential or difference equations that link the NN weights with the control input. When the output PDF is not measurable, Equation 1.5 should be used to obtain the output PDF from an input and output model. For example, tracking error entropy has been used in [188] as the closed-loop index so as to reduce the uncertainties and randomness embedded in the closed-loop tracking error. Indeed, initial developments of this scheme were reported in [188], where a recursive control algorithm has been developed that minimizes the tracking error entropy for the closed-loop system using Equation 1.6. The algorithm has been demonstrated via simulated examples and a comparison with the existing minimum variance control has also been made. To guarantee the closed-loop stability, the Youla parametrization technique has been applied to the initial structuring of the controller, where the freedom from the Youla parametrization is utilized in the entropy minimization.

Recently, the entropy optimization principle is firstly presented in [67, 69], a new auxiliary mapping from the input PDFs to output PDFs are constructed and optimization algorithms with stability analysis have been provided. On the other hand, the statistic tracking control or statistic information set control approaches will be presented in this book (see [68, 181, 183]).

## 1.5 Stochastic Distribution Filtering Design

Another important fields is on filter design for non-Gaussian stochastic systems (see [66, 68]). For example, in the well-known Kalman filtering algorithm the filter design objective is to realize a minimum variance residual. This has always been based on the Gaussian noise and linear system assumptions. When non-Gaussian, stochastic systems are considered, the minimization of the variance for the residual will not generally produce the required filtering effect. In this context, the filter design objective should be so as to realize a narrowly distributed Gaussian residual signal or to minimize the residual's randomness. The former is in fact a PDF shape control problem where the filter design problem is to select the filter parameters (i.e., the gain matrix of filters) so that the residual PDF is made to follow a narrowly distributed Gaussian PDF (see Figure 1.9). The later can be classified as a minimum entropy problem where the entropy of the filter's residual should be minimized during the filter design by selecting appropriate filter gains.

Another important problem is the use of the measured output PDFs to detect and diagnose a fault. In this respect, another new filtering algorithm has been developed, where the system is described by nonlinear difference equations (see [67]). A filtering system was then proposed which led to the error dynamics of the filtering algorithm. By establishing the concept of hybrid random variables and distributions, a new formulation of the residual PDF has been made so as to link the residual PDF to the filtering gain. As such, an optimal gain matrix was designed so that the entropy of the residual is minimized. Convergence analysis was also performed and several necessary convergence conditions were derived. In comparison with existing filtering algorithms, the developed minimum entropy filter has shown better performance in reducing the randomness of the filter's residual and is more general and therefore suited to non-Gaussian cases.

The novel research framework for the filtering, fault detection and diagnosis of stochastic distribution systems has been established in [66–69, 73, 196].

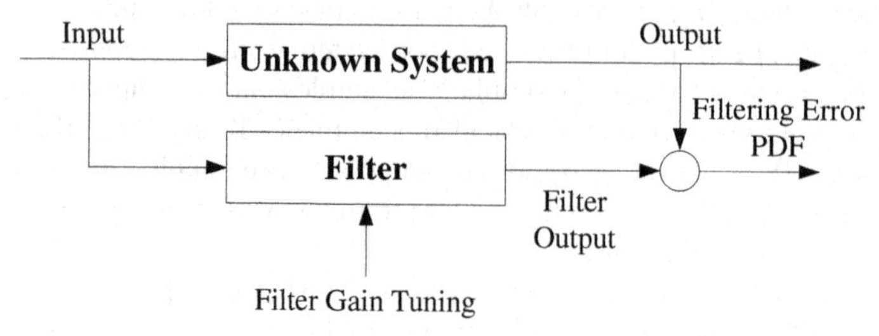


**Figure 1.9** Filtering design transferred into a PDF control problem

## 1.6 Conclusions

This chapter has reviewed the recently developments in SDC, which can also be regarded as output PDF control for non-Gaussian stochastic systems. Although the theory was motivated when designing controllers for a number of practical systems, it has found a much wider application in terms of filter design, data-driven modeling and data dimension reduction. In this context, it is expected that that SDC will have a wide range of applications in the near future.

Indeed, the theory described in previous sections has been successfully applied to a number of practical processes. The first application was made to paper making systems where the 2D solid distribution control of paper web was investigated. In this application, the pilot paper machine of the University of Manchester was used to develop the distribution control algorithms and both the output PDF shape and the entropy were controlled [61, 157, 188]. At present, an industrial scale system is being studied at Iggesund Paper Board plc in Workington (UK). The second application was made to MWD control and both B-spline-based and iterative learning-based output PDFs control were developed and applied to the systems. Desired output PDF tracking has been realized. The third application was performed for flame distribution control of a simple combustion system [146]. Moreover, the application of SDC in chemical engineering problems have also been made. These applications have shown clearly the advantages of using SDC design in real-time control.

# Part I
# Structural Controller Design for Stochastic Distribution Control Systems

As we discussed in Chapter 1, stochastic control has been an important research subject over the past decades. Many successful approaches have been developed including minimum variance control [3, 5], self-tuning control [5, 53], stochastic linear quadratic control [1, 3], control for stochastic NN [54, 175] and regulation for jumping parameter systems [7, 180]. A common feature in these works is that all variables in the system are supposed to obey a Gaussian distribution and only the output mean and variance are controlled. Motivated by several typical examples in practical systems such as paper and board making, a new group of control strategies for the output PDFs have been developed for both static and dynamic stochastic systems [61, 157, 164]. Different from any other previous stochastic control approaches, the variables can be non-Gaussian and the concerned output is in fact the whole shape of the output PDF. The objective of the controller design is therefore to find a crisp control input, namely the PDF control, so that the shape of output PDF can follow a given distribution.

For a general nonlinear stochastic system, it is difficult to determine its output PDFs using analytical methods due to the high nonlinearity of both the system and its PDFs. For example, even for a system described by an Ito's differential equation with a Wiener input, it is required that the PDFs of the state should satisfy a PDE named as the Fokker-Planck equation [3, 162]. As a result, in order to obtain some feasible design algorithms, B-spline expansions (see, e.g., [10]) have been introduced to model the dynamical relationships between the control input and the output PDFs [61, 63, 64, 157, 161–164, 166, 169]. However, it is noted that hitherto, the developed PDF control laws were obtained directly via numeral optimization processes. First, this may lead to difficulties in the realization of the complicated controllers. Second, since the numerical solution does not always generate a fixed closed-loop structure, the analysis of the closed-loop properties including stability and robustness is difficult to perform. Hence, it is necessary to develop output PDF control strategies that have a fixed closed-loop structure. Since in this case the controller contains a set of parameters, its design can be performed by optimizing these parameters.

Indeed, PI/PID control has been widely applied in many practical systems. Its design can be carried out based on both the frequency-domain and time-domain framework (see [2, 4, 47, 79, 91, 150] and references therein), where most results were devoted to linear SISO systems. For example, the approach proposed in [47] provided a convex solution for the PID-based LQG control problem of SISO systems by transforming the PID control function to a pure proportional control. For MIMO systems, it was shown that the procedure results in a static output feedback control with a special controller structure. Under some conditions, bilinear or iterative LMI algorithms were provided in [106, 116, 198]. It is noted that in [106, 116, 198], a non-singular constraint related to well-posedness should be guaranteed, which might result in non-convex algorithms and conservative results.

As a result, it is necessary to develop simple structured controllers for output PDF control. This forms the main objective of this part, where a PDF control algorithm with a fixed structure, namely the pseudo-PI/PID structural controller based on LMIs, is descried in the following.

# Chapter 2
# Proportional Integral Derivative Control for Continuous-time Stochastic Systems

## 2.1 Introduction

This chapter presents a pseudo-proportional integral derivative (PID) tracking control strategy for general non-Gaussian stochastic systems based on a linear B-spline model for the output PDFs. The objective is to control the conditional PDFs of the system output to follow a given target function. Different from existing methods, the control structure (i.e., the PID) is imposed before the output PDF controller design. Following the linear B-spline approximation on the measured output PDFs, the problem is transferred into the tracking of given weights which correspond to the desired PDF. In this case, since the output is the shape of the PDFs of the random output variables of systems, the classical PID controllers cannot be applied directly. Instead, a pseudo-PID control strategy will be presented for the weight tracking control problem, for which a dynamical model is established to describe the relationship between the weight and the output PDFs. For systems with or without model uncertainties, it is shown that the solvability can be cast into a group of matrix inequalities. Furthermore, an improved controller design procedure based on the convex optimization is proposed, which can guarantee the required tracking convergence with enhanced robustness. Simulations are given to demonstrate the efficiency of the proposed approach and encouraging results have been obtained.

For a general stochastic system, it is supposed that $u(t)$ and $z(t)$ are the control input and system output, respectively. The probability of the random variable $z(t)$ lying inside $[a,y]$ is denoted by $P(a \leq z(t) < y, u(t))$, and the corresponding conditional PDF is denoted by $\gamma(y,u(t))$, where $y$ represents the corresponding sample variable.

For a function $\gamma(y,u(t))$, B-spline expansions can be used to approximate it. It is denoted that $C(y) = [c_1(y), \cdots, c_{n-1}(y)]$ represents the pre-specified independent basis function vector and $v(t) = [v_1(t), \cdots, v_{n-1}(t)]$ is the corresponding dynamical weight in such a B-spline approximation. $g(y)$ is the target PDF to be tracked and $V_g$ is the desired weighting corresponding to $g(y)$. $e(y,u(t))$ is used to denote the difference between $\gamma(y,u(t))$ and $g(y)$.

## 2.2 Problem Formulation

### 2.2.1 Model Representation

In practice, the system input may be non-Gaussian, which will result in non-Gaussian output. In fact, when the system concerned is nonlinear, the system output may also be a non-Gaussian variable. For such general stochastic systems, the mean and variance of the output may be insufficient to characterize the precise properties of the system output. This is the main difference from the classical stochastic control approaches where only mean or variance are concerned. On the other hand, in many practical processes, only some statistical information can be measured and obtained, rather than the value of some variables. This feature also requires a new type of stochastic control strategy. In the following, B-spline expansions will be adopted to model the output PDFs.

Consider Figure 2.1, which represents a general stochastic system, where $\mathrm{d}(t)$ is the stochastic input, $u(t) \in R^m$ is the control input. It is supposed that $z(t) \in [a,b]$ is the system output and the probability of output $z(t)$ lying inside $[a,y]$ can be described as

$$P(a \leq z(t) < y, u(t)) = \int_a^y \gamma(\eta, u(t))\mathrm{d}\eta$$

where $\gamma(y,u(t))$ is the PDF of the stochastic variable $y(t)$ under control input $u(t)$. It is supposed in this chapter that the output PDF $\gamma(y,u(t))$ as the control objective, can be measured or estimated (see [157] and references therein). For example, in the paper making process, $u(t)$ can be regarded as a retention polymer flow rate, and $y(t)$ can represent the pore size of the fibrous network in the wire section of a paper machine as a non-Gaussian random variable, whose distributions are measurable on-line by several sensors [164]. The above system is multiple input and single output (MISO). However, for multiple output systems, combined output PDFs can be used, where the shape control would be addressed to the combined output PDF.

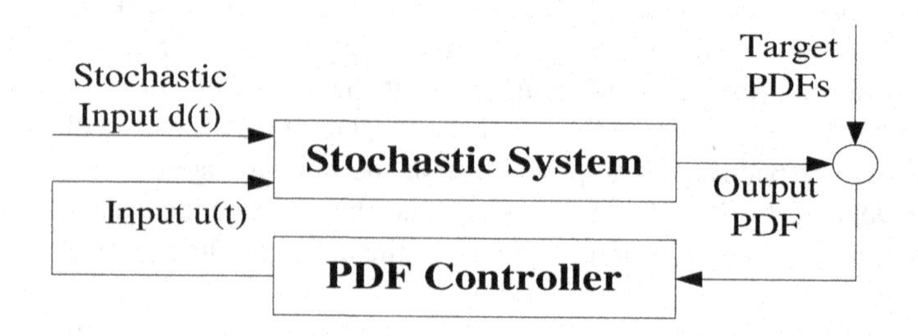


**Figure 2.1** PDF tracking control for a stochastic system using measured PDFs

To avoid the complex computation modeling involved in PDEs and to provide crisp control strategies, new modeling procedures for PDF control problems related to the linear B-spline approximation have been presented, where $\gamma(y,u(t))$ can be represented by

$$\gamma(y,u(t)) = C(y)v(t) + L(y), \ \ y \in [a,b], \tag{2.1}$$

and the column weight vector $v(t) \in R^n$ satisfies

$$\dot{v}(t) = Av(t) + Bu(t). \tag{2.2}$$

In Equations 2.1 and 2.2, $C(y)$ is the pre-specified basis function row vector, which is related to B-spline expansions. It is noted that generally Equations 2.1 and 2.2 should contain both the approximation errors and the modeling uncertainties (see [164]), this case will also be studied in this chapter after the problem for nominal models (2.1) and (2.2) is solved.

The desired PDF for $\gamma(y,u(t))$ to follow is assumed to be

$$g(y) = C(y)v_g + L(y) \tag{2.3}$$

where $v_g$ is the desired weight with respect to the same $C(y)$. As described in Chapter 1, $g(y)$ is in fact the set-point for the output PDF to follow, and is determined by the actual requirement.

Using these definitions, the tracking error $e(y,t)$ can be described as

$$e(y,t) = g(y) - \gamma(y,u(t)) = C(y)w(t), \tag{2.4}$$

which is a function of both $y \in [a,b]$ and the time instant $t$, where the weight error vector is defined as

$$w(t) = v_g - v(t).$$

It is noted that in [163, 164], the design approaches were obtained by optimizing a performance index involved in $w(t)$. Thus, due to the complexity of the controller structure, neither stability of the closed-loop systems nor the asymptotic convergence of the tracking error can be guaranteed by most of those optimal controllers.

Based on the continuity theory of functions, it is noted that $e(y,t) \to 0$ if and only if $w(t) \to 0$. As a result, after establishing dynamic models (2.1) and (2.2) which combine the output PDFs with the control input through the weight vector, the aim of the controller design in this chapter can be transferred into finding the control input such that tracking with respect to the weight vectors is realized.

### 2.2.2 Pseudo-PID Controller Structure

To solve the tracking problem for PDF control, a direct application of the classical PID control strategy would lead to

$$u_g(y,t) = K_P e(y,t) + K_I \int_0^t e(y,\tau)d\tau + K_D \frac{\partial e(y,t)}{\partial t} \tag{2.5}$$

where $K_P, K_I$ and $K_D$ are control gains to be determined. However, since $y$ is a variable defined on $[a,b]$, $u_g(y,t)$ is a twin-variable function; such a control action is not realizable in practice since stochastic systems can only accept one input value (or vector) at a time instant. To overcome this obstacle, instead of Equation 2.5, we focus on the pseudo-PID control law constructed as follows

$$u = [K_P \int_a^b C(y)\Gamma_p(y)\mathrm{d}y]w + [K_I \int_a^b C(y)\Gamma_I(y)\mathrm{d}y] \int_0^t w\mathrm{d}\tau$$

$$+[K_D \int_a^b C(y)\Gamma_D(y)\mathrm{d}y]\dot{w} \tag{2.6}$$

where $\Gamma_p(y), \Gamma_I(y)$ and $\Gamma_D(y)$ are weight diagonal square matrices to be designed. Denote

$$\begin{cases} \bar{K}_P = [K_P \int_a^b C(y)\Gamma_P(y)\mathrm{d}y] \\ \bar{K}_I = [K_I \int_a^b C(y)\Gamma_I(y)\mathrm{d}y] \\ \bar{K}_D = [K_D \int_a^b C(y)\Gamma_D(y)\mathrm{d}y] \end{cases} . \tag{2.7}$$

It is noted that if $K_P, K_I$ and $K_D$ are $m$ dimension column vectors, $\overline{K}_P, \overline{K}_I$ and $\overline{K}_D$ should be $m \times n$ matrices. Following this transformation, the problem can be simplified and transformed to one of finding $\overline{K}_P, \overline{K}_I$ and $\overline{K}_D$ such that $w(t)$ converges to zero.

*Remark 2.1* If the parameter estimation algorithm involved in training the weight $v(t)$ has good computational accuracy and speed, then the control law

$$u = \overline{K}_P w + \overline{K}_I \int_0^t w\mathrm{d}\tau + \overline{K}_D \dot{w} \tag{2.8}$$

can be applied directly. Otherwise, we should still solve $K_P, K_I$ and $K_D$ for a group of $\Gamma_p(y), \Gamma_I(y)$ and $\Gamma_D(y)$ by using Equation 2.7 under proper rank conditions. In this case, Equation 2.6 will be used instead of Equation 2.8.

*Remark 2.2* If the output of the system (2.2) is represented by $z = Hv$, the pseudo-PID control law can be given by

$$u = \overline{K}_P z + \overline{K}_I \int_0^t z\mathrm{d}\tau + \overline{K}_D \dot{z} .$$

It should be pointed out that this more general case can also be considered similarly.

## 2.3 Pseudo-PID Controller Design

### 2.3.1 Solvability Condition

For simplicity, we denote

$$s = \int_0^t w(\tau)\mathrm{d}\tau, \;\; \Pi = (A - B\overline{K}_P), \;\; \Delta = I + B\overline{K}_D. \tag{2.9}$$

Substituting Equation 2.8 into Equation 2.2 yields the following closed-loop system equations:

$$\begin{bmatrix} \dot{w} \\ \dot{s} \end{bmatrix} = \begin{bmatrix} \Delta^{-1}\Pi & -\Delta^{-1}B\overline{K}_I \\ I & 0 \end{bmatrix} \begin{bmatrix} w \\ s \end{bmatrix} + \begin{bmatrix} -\Delta^{-1}A \\ 0 \end{bmatrix} v_g. \tag{2.10}$$

As a result, the objective is transferred into finding a set of $\overline{K}_P, \overline{K}_I$ and $\overline{K}_D$ such that system (2.10) is stable and $w(t)$ is convergent.

To formulate the proposed control algorithm, $\overline{K}_D$ should be demanded such that $\Delta$ is invertible. In this case, the singular situation (see [60]) can be avoided. Different from the approaches of [106, 116, 198], the PID tracking control problem other than a pure stabilization problem is considered in the chapter, and the closed-loop system (2.10) has a new style and lower dimensions.

**Theorem 2.1** The closed-loop system combined with the weight dynamical system (2.2) and the pseudo-PID control law (2.8) is asymptotically stable if the following two algebraic matrix inequalities

$$P = \begin{bmatrix} P_1 & P_2 \\ P_2^T & P_3 \end{bmatrix} > 0, \tag{2.11}$$

$$\Phi = \begin{bmatrix} \Phi_{11} & \Phi_{12} \\ \Phi_{12}^T & \Phi_{22} \end{bmatrix} < 0 \tag{2.12}$$

are solvable for $P > 0$ and matrices $\overline{K}_P$, $\overline{K}_I$ and $\overline{K}_D$, where

$$\begin{aligned} \Phi_{11} &:= \mathrm{sym}(P_1\Delta^{-1}\Pi + P_2), \\ \Phi_{12} &:= -P_1\Delta^{-1}B\overline{K}_I + \Pi^T\Delta^{-T}P_2 + P_3, \\ \Phi_{22} &:= \mathrm{sym}(-P_2^T\Delta^{-1}B\overline{K}_I). \end{aligned}$$

$I + B\overline{K}_D$ is invertible. In this case, the output PDF $\gamma(y, u(t))$ converges to the desired PDF $g(y)$.

*Proof.* To prove this theorem, three targets should be achieved under conditions (2.11) and (2.12):

1. the states of system (2.10) are bounded for all $t$;

2. the system (2.10) is asymptotically stable;
3. $\lim_{t\to\infty} w(t) = 0$.

First, the Lyapunov stability theorem is applied directly to system (2.10) under conditions (2.11) and (2.12). For this purpose, the Lyapunov candidate can be selected as

$$V(w,s,t) = \begin{bmatrix} w^T & s^T \end{bmatrix} \begin{bmatrix} P_1 & P_2 \\ P_2^T & P_3 \end{bmatrix} \begin{bmatrix} w \\ s \end{bmatrix} \tag{2.13}$$

when condition (2.11) holds.

If condition (2.12) holds, there must exist a positive scalar $\sigma_1 > 0$ such that $\Phi < -\sigma_1 I$. Along with the trajectories of system (2.10) it can be verified that

$$\begin{aligned}
\dot{V}(w,s,t) &= \begin{bmatrix} w^T & s^T \end{bmatrix} P \begin{bmatrix} \dot{w} \\ \dot{s} \end{bmatrix} + \begin{bmatrix} \dot{w}^T & \dot{s}^T \end{bmatrix} P \begin{bmatrix} w \\ s \end{bmatrix} \\
&= \begin{bmatrix} w^T & s^T \end{bmatrix} \Phi \begin{bmatrix} w \\ s \end{bmatrix} + 2 \begin{bmatrix} w^T & s^T \end{bmatrix} \begin{bmatrix} -P_1 \Delta^{-1} A \\ -P_2^T \Delta^{-1} A \end{bmatrix} v_g \\
&< -\sigma_1 \left\| \begin{bmatrix} w \\ s \end{bmatrix} \right\|^2 + 2 \left\| \begin{bmatrix} w \\ s \end{bmatrix} \right\| \left\| \begin{bmatrix} -P_1 \Delta^{-1} A \\ -P_2^T \Delta^{-1} A \end{bmatrix} \right\| \|v_g\| .
\end{aligned}$$

Thus, $\dot{V}(w,s,t) < 0$ if

$$\left\| \begin{bmatrix} w \\ s \end{bmatrix} \right\| \geq 2\sigma_1^{-1} \left\| \begin{bmatrix} -P_1 \Delta^{-1} A \\ -P_2^T \Delta^{-1} A \end{bmatrix} \right\| \|v_g\| := \sigma_2,$$

which means that for all $t$ the state vector $\begin{bmatrix} w^T & s^T \end{bmatrix}^T$ satisfies

$$\left\| \begin{bmatrix} w(t) \\ s(t) \end{bmatrix} \right\| \leq \max \left\{ \left\| \begin{bmatrix} w(0) \\ s(0) \end{bmatrix} \right\|, \sigma_2 \right\}.$$

In the next phase of the proof, we show that there is a unique equilibrium point of system (2.10) for the given $v_g$. Suppose that $\theta_1(t)$ and $\theta_2(t)$ are two trajectories of the closed-loop system (2.10) corresponding to a fixed initial condition and the input $v_g$. Then the variable $\rho(t) = \theta_1(t) - \theta_2(t)$ satisfies the dynamic equation

$$\dot{\rho} = \begin{bmatrix} \Delta^{-1}\Pi & -\Delta^{-1} B \bar{K}_I \\ I & 0 \end{bmatrix} \rho \tag{2.14}$$

with initial condition $\rho(0) = 0$. It can be verified that $\rho(t) = 0$ is the unique asymptotically stable equilibrium point of the system (2.14) by using the same Lyapunov function (2.13). It means that the closed-loop system (2.10) also has a unique asymptotically stable equilibrium point.

Finally, it is left to verify that the tracking performance should be guaranteed. For this purpose, let us assume that $\theta(t)$ is a trajectory of the linear closed-loop system (2.10), and

$$\lim_{t\to\infty} \theta(t) = \theta_0$$

where $\theta_0$ is a constant vector; it follows that $\lim_{t\to\infty} \dot{\theta}(t) = 0$. Since from system (2.10) it can be seen that $\dot{s} = w(t)$, it can be concluded that $\lim_{t\to\infty} w(t) = 0$. □

### 2.3.2 *Feasible Design Procedures*

From Theorem 2.1, it can be seen that

$$\text{sym}(-P_2^T \Delta^{-1} B\overline{K}_I) < 0 \tag{2.15}$$

is a necessary condition of matrix inequality (2.12) and with a strong nonlinearity with respect to $P_2$, $\overline{K}_D$ and $\overline{K}_I$. The following lemma needs to be provided to deduce the nonlinearity.

**Lemma 2.1** If there exist a row vector $P_I$ and a matrix $P_D$ such that

$$\text{sym}(-BP_I - BP_D B^T) < 0, \tag{2.16}$$

then matrix inequality (2.15) holds for some $P_2$ and $\overline{K}_I$. In this case, $P_2$ and $\overline{K}_I$ can be obtained via

$$\overline{K}_I = P_I P_2 \tag{2.17}$$

and $\overline{K}_D$ can be solved via

$$P_I \overline{K}_D^T = P_D \tag{2.18}$$

such that the obtained $\Delta$ is also invertible. Alternatively, we choose $P_2 = \frac{1}{2}\alpha I > 0$ in Equation 2.17 for an appropriate $\alpha > 0$.

*Proof.* Matrix inequality (2.15) is equivalent to

$$\text{sym}(-B\overline{K}_I P_2^{-1} \Delta^T) < 0 \tag{2.19}$$

by pre-multiple $\Delta P_2^{-T}$ and post-multiple $P_2^{-1}\Delta^T$.

After denoting

$$P_I = \overline{K}_I P_2^{-1}, P_D = \overline{K}_I P_2^{-1} \overline{K}_D^T \tag{2.20}$$

it can be verified that matrix inequality (2.19) holds if and only if matrix inequality (2.16) holds. Obviously, matrix inequality (2.16) is an LMI with respect to $P_I$ and $P_D$. Thus a group of $P_2$, $\overline{K}_I$ and $\overline{K}_D$ can be computed, where $\overline{K}_D$ can guarantee that $\Delta$ is invertible. At this stage, it is easy to compute $\alpha$ such that $\alpha > 0$ and $P_2 + P_2^T \leq \alpha I$. To simplify this procedure, a possible choice of $P_2 = \frac{1}{2}\alpha I > 0$ can be made for any given $\alpha > 0$. □

With Lemma 2.1, the following simplified result can be proposed to provide an improved algorithm.

**Theorem 2.2** For the $\alpha$, $\overline{K}_I$ and $\overline{K}_D$ obtained in Lemma 2.1, if there exist matrices $Q_1 > 0$, $Q_P$ and a parameter $\delta > 0$ satisfying

$$\Psi = \begin{bmatrix} \Psi_{11} & \Psi_{12} + \delta^2 I & Q_1 \\ \Psi_{12}^T + \delta^2 I & \text{sym}(-\alpha\Delta^{-1}B\overline{K}_I) & 0 \\ Q_1^T & 0 & -\alpha^{-1}I \end{bmatrix} < 0 \tag{2.21}$$

where

$$\Psi_{11} := \text{sym}(\Delta^{-1}AQ_1^T - \Delta^{-1}BQ_P^T),$$
$$\Psi_{12} := -\Delta^{-1}B\overline{K}_I + \alpha(Q_1A^T - Q_PB^T)\Delta^{-T}$$

with the constraint

$$\frac{\alpha}{2}\lambda\text{max}(Q_1) < \delta, \tag{2.22}$$

then the closed-loop system combined with the weight dynamical system (2.2) and the pseudo-PID control law (2.8) is asymptotically stable and the output PDF $\gamma(y,u)$ is convergent to the desired PDF $g(y)$. In this case, $\overline{K}_P$ can be computed via $\overline{K}_P = Q_P^T Q_1^{-T}$.

*Proof.* In this proof, it should be shown that if matrix inequality (2.21) holds for some $Q_1 > 0$, $Q_P$ and $\delta > 0$ along with the obtained $\alpha$, $\overline{K}_I$ and $\overline{K}_D$ in Lemma 2.1, then there must exist $P > 0$ and a group of $\overline{K}_P$, $\overline{K}_I$ and $\overline{K}_D$ satisfying both conditions (2.11) and (2.12).

Denote

$$Q_1 = P_1^{-1}, \Omega_1 = \text{diag}\{P_1^{-1}, I\}. \tag{2.23}$$

Substituting the obtained $P_2$, $\overline{K}_I$ and $\overline{K}_D$ into matrix inequality (2.12), and pre-multiplying $\Omega_1$ and post-multiplying $\Omega_1^T$ yields

$$\begin{bmatrix} \bar{\Psi}_{11} & \bar{\Psi}_{12} \\ \bar{\Psi}_{12}^T & \bar{\Psi}_{22} \end{bmatrix} < 0 \tag{2.24}$$

where

$$\bar{\Psi}_{11} := \text{sym}(\Delta^{-1}AQ_1^T - \Delta^{-1}BQ_P^T) + Q_1(P_2 + P_2^T)Q_1^T,$$
$$\bar{\Psi}_{12} := -\Delta^{-1}B\overline{K}_I + (Q_1A^T - Q_PB^T)\Delta^{-T}P_2 + Q_1P_3,$$
$$\bar{\Psi}_{22} := \text{sym}(-P_2^T\Delta^{-1}B\overline{K}_I)].$$

Matrix inequality(2.24) is equivalent to

$$\begin{bmatrix} \Psi_{11} & \Psi_{12}+Q_1P_3 & Q_1 \\ \Psi_{12}^T+P_3^TQ_1^T & \text{sym}(-P_2^T\Delta^{-1}B\overline{K}_I) & 0 \\ Q_1^T & 0 & -\alpha^{-1}I \end{bmatrix} < 0 \tag{2.25}$$

by using the well-known Schur complement formula.

Denote $Q_3 = Q_1P_3 = \delta^2 I$. It is obvious that matrix inequality (2.25) is equivalent to matrix inequality (2.21) when $P_3 = \delta^2 Q_1^{-1}$. Finally, using Schur complement, condition (2.11) holds if

$$P_3 - \frac{1}{4}\alpha^2 P_1^{-1} > 0$$

or

$$(\delta^2 Q_1^{-2} - \frac{1}{4}\alpha^2)Q_1 > 0.$$

Indeed, these inequalities can be guaranteed by matrix inequality (2.22). □

*Remark 2.3* Similarly to previous work on PID controller design (see [4, 79, 106, 116, 150, 198]), Theorem 2.1 depends on the solution of a non-convex problem due to nonlinear algebraic matrix inequality (2.12). Alternatively, an iterative algorithm can be applied to calculate $P_I$, $\alpha_D$, $\delta$, $Q_1$ and $Q_P$ by using conditions (2.11) and (2.12). In Theorem 2.2 a simplified algorithm based on convex computation is presented through some proper transformations and decompositions on condition (2.12).

*Remark 2.4* Compared with previous work on PID control of MIMO systems [106, 116, 198], there are several features of our results for the transformed PID weight tracking control problem. First, it is noted that tracking control is discussed in this chapter rather than a pure stabilization problem. Next, in Theorems 2.1 and 2.2, the well-posedness problem can be avoided naturally by constructing $\overline{K}_D$ via a linear equation $P_I\overline{K}_D^T = \alpha_D$, and the conservative additive condition (see (14) in [116]) to guarantee well-posedness can be removed. Another feature of the presented approach is that a full block form for the Lyapunov matrix corresponding to the Lyapunov function (2.13) is chosen here, which is less conservative than those cases where only diagonal Lyapunov matrices were considered (see [116]). Finally, it is shown that the PID tracking controller design method presented in Theorem 2.2 is based on a convex optimization while in [106, 116, 198] PID control was cast into a static output feedback control framework, which leads to a non-convex optimization problem in the LMI context.

*Remark 2.5* Using the LMI-toolbox, a feasible design algorithm can be summarized as follows:

*Step 1: Solve LMI (2.16) for the row vector $P_I$ and scalar $\alpha_D$.*
*Step 2: Solve algebraic equation $\overline{K}_I = P_I P_2$ to obtain $P_2 = \frac{1}{2}\alpha I$ and $\overline{K}_I$, and $P_I\overline{K}_D^T = \alpha_D$ to get $\overline{K}_D$ such that $\Delta$ is invertible.*
*Step 3: For a given $\delta > 0$, solve LMIs (2.21) and (2.22) for $Q_1 > 0$, $Q_P$.*
*Step 4: Compute $\overline{K}_P$ through $\overline{K}_P = Q_P^T Q_1^{-T}$ and construct the controller as control law (2.8).*

If the pseudo-PID control law (2.5) is required, $K_P, K_I$ and $K_D$ should be constructed based on Equation 2.7 for a group of $\Gamma_P(y), \Gamma_I(y)$ and $\Gamma_D(y)$ using $\overline{K}_P$, $\overline{K}_I$ and $\overline{K}_D$. This case requires that $P_D$ and $Q_P$ are rank 1 matrices and may lead to some iterative algorithms [8].

### *2.3.3 Robust Pseudo-PID Controller*

In practical processes, B-spline approximation may result in some uncertainties so that the precise model (2.1) is unavailable. In [164], conditions on the closed-loop stability and robustness analysis were presented for the PDF control problem using the optimization approach, which are unfortunately difficult to apply to controller design. In this section, the results for nominal systems (2.1) and (2.2) given in Theorems 2.1 and 2.2 will be generalized to the uncertain case.

In this case, instead of model (2.1), the following B-spline expansion

$$\gamma(y,u) = C(y)v + L(y) + \eta(y,u) \tag{2.26}$$

should be considered, where $\eta(y,u(t))$ can be considered as an unknown, smooth and bounded function.

Also, both un-modeled dynamics and parametric uncertainty can be included in the weight dynamics (2.2). As a result, the following weight dynamic model

$$\dot{v}(t) = Av(t) + Bu(t) + d_1(t) \tag{2.27}$$

should be used instead of model (2.2), where $A$ is an unknown matrix belonging to a set described by

$$\left\{ \sum_{i=1}^{N} \lambda_i A_i, \sum_{i=1}^{N} \lambda_i = 1 \right\},$$

and $A_i$ are given extreme points of the polyhedral. $d_1(t)$ can be regarded as a computational error or an external disturbance in this case.

To simplify the description of $\eta(y,u)$, we assume that $\eta(y,u) = C(y)\xi(t)$, where $\xi(t)$ can be regarded as the system response of model (2.27) with respect to an unknown additional input $d_2(t)$. Thus, both the uncertainties $\eta(y,u)$ and $d_1(t)$ can be summarized into the following model:

$$\dot{v}(t) = Av(t) + Bu(t) + d(t) \tag{2.28}$$

where $d(t) = d_1(t) + d_2(t)$ is assumed to be bounded.

After substituting Equation 2.6 into Equation 2.28, it can be shown that the closed-loop system is given by

$$\begin{bmatrix} \dot{w} \\ \dot{s} \end{bmatrix} = \begin{bmatrix} \Delta^{-1}\Pi & -\Delta^{-1}B\overline{K}_I \\ I & 0 \end{bmatrix} \begin{bmatrix} w \\ s \end{bmatrix} + \begin{bmatrix} -\Delta^{-1}A & \Delta^{-1} \\ 0 & 0 \end{bmatrix} \begin{bmatrix} v_g \\ d \end{bmatrix}. \tag{2.29}$$

For this system, its stability and tracking performance can be summarized by the following theorem.

**Theorem 2.3** For the $\alpha$, $\overline{K}_I$ and $\overline{K}_D$ obtained in Lemma 2.1, if there exist matrices $Q_1 > 0$, $Q_P$ and a parameter $\delta > 0$ satisfying

$$\Psi_i = \begin{bmatrix} \Psi_{11i} & \Psi_{12i}+\delta I & Q_1 \\ \Psi_{12i}^T+\delta I & \mathrm{sym}(-\alpha\Delta^{-1}B\overline{K}_I) & 0 \\ Q_1^T & 0 & -\alpha^{-1}I \end{bmatrix} < 0, \tag{2.30}$$

and the constraint (2.22) for $i = 1,2,\cdots N$, where

$$\Psi_{11i} := \mathrm{sym}(\Delta^{-1}A_iQ_1^T - \Delta^{-1}BQ_P^T),$$

$$\Psi_{12i} := -\Delta^{-1}B\overline{K}_I + \alpha(Q_1A_i^T - Q_PB^T)\Delta^{-T},$$

then the closed-loop system combined with the weight dynamical system (2.28) and the pseudo-PID control law (2.6) is robustly asymptotically stable and the output PDF $\gamma(y,u)$ is convergent to the desired PDF $g(y)$. In this case, $\overline{K}_P$ can be computed via $\overline{K}_P = Q_P^TQ_1^{-T}$.

*Proof.* Theorem 2.2 still holds for the system (2.29), except that in system (2.29) the coefficient matrices may be polytopes rather than known ones. It can be observed that $A$ is linear in LMI (2.21). Thus, LMI (2.21) holds for the whole polytope if and only if it does for its all extreme points. The rest of proof is similar to the proof of Theorem 2.1 and 2.2. It can be shown that system (2.29) is stable as long as $d(t)$ is bounded. □

The design procedures for the robust pseudo-PID tracking controllers can be provided similarly to Remark 2.5, where LMI (2.21) should be replaced by a group of LMIs described by matrix inequality (2.30).

## 2.4 Simulations

For a stochastic system with non-Gaussian process, it is supposed that the output PDF can be formulated to be Equation 2.1 with $v = \begin{bmatrix} v_1 & v_2 \end{bmatrix}^T$ and

$$B_i(y) = \begin{cases} |\sin 2\pi y|, y \in [0.5(i-1), 0.5i], i = 1,2,3 \\ 0, \quad otherwise \end{cases},$$

$$C(y) = \begin{bmatrix} B_1(y) - B_3(y) & B_2(y) - B_3(y) \end{bmatrix},$$

$$L(y) = \frac{B_3(y)}{\int_0^{1.5} B_3(y)dy},$$

for $i = 1,2,3$, $y \in [0,1.5]$, after a B-spline approximating procedure.

The dynamical relations between $v$ and $u$ are described by Equation 2.2 with

$$A = \begin{bmatrix} a_{11} & -1 \\ 0 & a_{22} \end{bmatrix}, B = \begin{bmatrix} 1 & 0 \\ 0 & 2 \end{bmatrix},$$

where $a_{11} \in [1.8, 2.2]$ and $a_{22} \in [0.9, 1.1]$.

The desired PDF $g(y)$ is supposed to be described by Equation 2.3 with $v_g = \left[\frac{\pi}{5}\ \frac{\pi}{4}\right]^T$. In the simulation, the initial PDF is given by Equation 2.1 with $v(0) = \left[\frac{\pi}{3}\ \frac{\pi}{3}\right]^T$. Corresponding to model (2.28), $d(t)$ is supposed to be a random signal with amplitude 0.2 and frequency 1Hz.

First, it can be found via LMI (2.16) that

$$P_I = \begin{bmatrix} 8.1614 & 0 \\ 0 & 6.2449 \end{bmatrix}, P_D = \begin{bmatrix} 1 & 0 \\ 0 & 1 \end{bmatrix},$$

with which it can be obtained that

$$P_2 = \begin{bmatrix} 1 & 0 \\ 0 & 1 \end{bmatrix}, \overline{K}_I = \begin{bmatrix} 8.1614 & 0 \\ 0 & 6.2449 \end{bmatrix}, \overline{K}_D = \begin{bmatrix} 0.1225 & 0 \\ 0 & 0.1601 \end{bmatrix}, \alpha = 2.$$

Solving LMI (2.21) produces $\delta^2 = 17.8711$ and

$$Q_1 = \begin{bmatrix} 2.0264 & 0 \\ 0 & 1.7670 \end{bmatrix}, Q_P = \begin{bmatrix} 14.5740 & 0 \\ -1.7670 & 6.1642 \end{bmatrix}.$$

Thus, it can be obtained that

$$\overline{K}_P = \begin{bmatrix} 7.1920 & -1 \\ 0 & 3.4885 \end{bmatrix}.$$

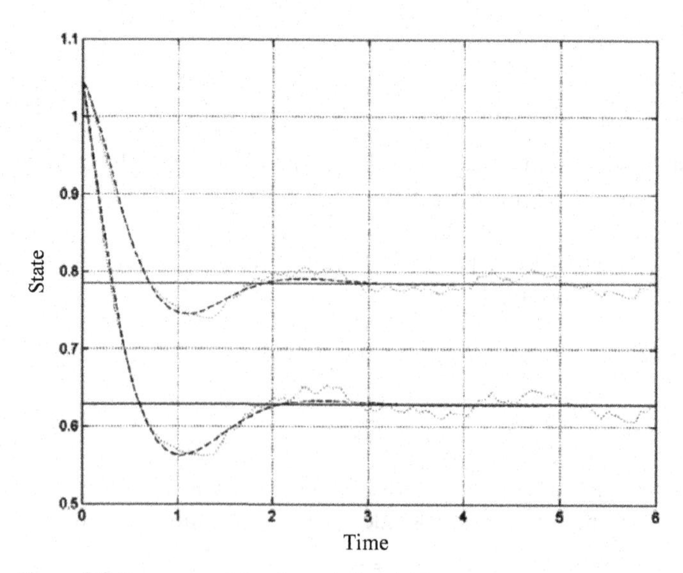


**Figure 2.2** Responses of the dynamical weighing systems

When the error of B-spline expansions is considered and reduced to the uncertain weighting system described by model (2.27), the PDF controller is termed a robust

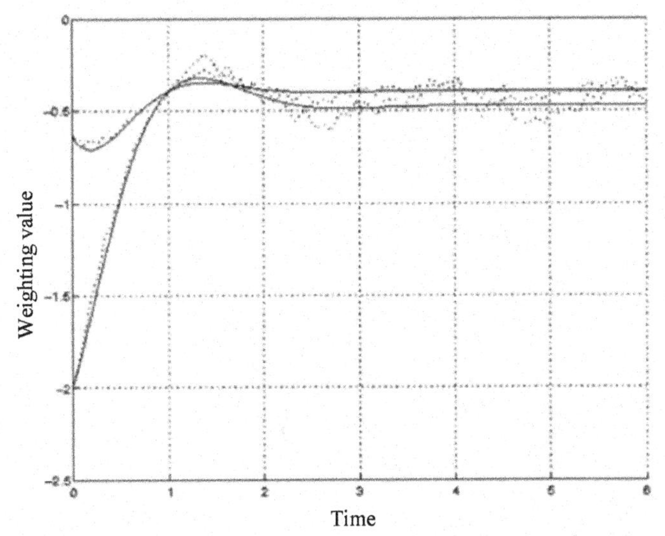


**Figure 2.3** Gains of the tracking controller for the weighting system with and without uncertainties

PDF controller. Correspondingly, the robust pseudo-PID controller can be obtained similarly by using Theorem 2.3. When the pseudo-PID control law and the robust one are applied, the closed-loop system responses for the dynamical weighting are shown in Figure 2.2. In Figure 2.2, the solid line represents the reference weight, the dashed line stands for the system response without disturbance $d(t)$ or uncertainty, and the dotted line represents the response with $d(t)$ or uncertainty. The control gains are shown in Figure 2.3, where the dotted line and the solid line correspond to the cases with and without system uncertainties, respectively. The practical PDFs for the uncertain weighting system and under the proposed robust control strategy is shown in Figure 2.4. It is demonstrated that satisfactory tracking performance and robustness are achieved.

## 2.5 Conclusions

A pseudo-PID tracking control law with respect to PDF control is first proposed for a general non-Gaussian stochastic systems. After B-spline approximation to the measured output PDFs, the control objective is transferred into the tracking of a group of given weights that correspond to the desired PDF. For both deterministic and uncertain cases, the solvability conditions are obtained in terms of a group of matrix inequalities for the pseudo-PID tracking control problem. Feasible controller design procedures are provided to guarantee the closed-loop stability and the tracking convergence. Different from the existing results on PDF control, the control strategy proposed in this chapter has a simple fixed structure and can guarantee both stability and robustness of the closed-loop system. Even compared with the results

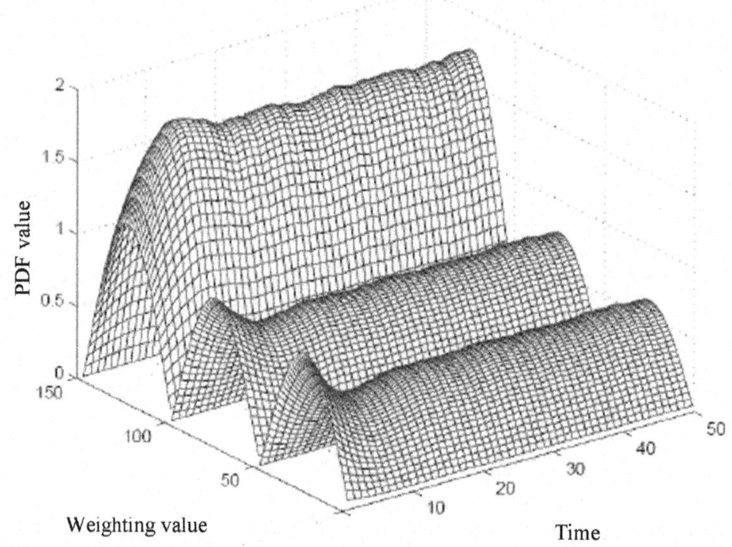


**Figure 2.4** 3D mesh plot of the output PDFs with the robust PDF controller

on the classical PID control problem of MIMO linear systems [116, 198], there are several features of the proposed results (see Remark 2.4). Simulation results show that satisfactory robustness and closed-loop performance can be achieved.

# Chapter 3
# Constrained Continuous-time Proportional Integral Derivative Control Based on Convex Algorithms

## 3.1 Introduction

It is noted that the tracking problem for the output PDF can be considered as an infinite dimensional tracking problem for a function. Up to now, most of the proposed PDF tracking control laws were obtained through the numeral optimization techniques (see [157, 158, 160, 163, 164]). As a result, the control structures were complicated and the rigorous stability analysis was difficult to provide. Since the controlled output is the shape of the conditional output PDF, the B-spline NN approximation has been effectively used for the modeling of output PDFs so that the problem can be reduced to a tracking problem for the weight systems [58, 61, 63, 64, 88, 160, 161, 163, 164, 166, 167, 173, 181–184, 188, 196]. It is noted that for the square root B-spline expansion model, a constraint on the weight vector should be satisfied, which leads to some additional obstacles in the tracking controller design. Generally speaking, three tasks should be achieved in the PDF tracking problem.

First, either the B-spline expansions or weighting modeling procedures may result in nonlinear dynamics, modeling errors and uncertainties, which have been ignored in most cases. Actually, errors exist in such types of neurofuzzy modeling processes, which may also be amplified in the corresponding tracking control procedures. This is because in the PDF tracking formulation, the output PDF and its B-spline model will be used for feedback as the measurement. In the previous chapter, a bound has been given to show how the modeling error influences the tracking error. However, analysis results were provided there which were difficult to use for controller design. The systems studied in [63, 64, 88, 157, 160, 161] and the previous chapter are linear models, where the shape of PDFs cannot be changed.

The second point is that the constraint required by the characteristics of the PDF were neglected in most of the existing results. It is noted that without such a constraint, the weighting vector is irrelevant to a PDF via B-spline models. In this case, the tracking objective of the output PDF cannot be achieved even if the weighting tracking is completed.

Finally, it is shown that the design algorithms provided to some extent are conservative and are complicated in computation. In particular, when the constraints are considered, the tracking problem turns into an invariant set control context with a non-zero equilibrium. For example, in Chapter 2, the precise B-spline expansion is used and the weighting model is formulated as a linear system in the absence of disturbance and uncertainties. Moreover, the algorithms in Theorem 2.1 of Chapter 2 involve nonlinear matrix inequalities and the result in Theorem 2.2 leads to conservative solutions.

To overcome the above obstacles, in this chapter we consider the generalized PID control for the SDC problem by using square root B-spline models and nonlinear weighting models with uncertainty and disturbances. The objective is to control the PDF of the system output to follow a given target function. Using the B-spline expansion and the nonlinear weighting model with exogenous disturbances, PDF tracking is transformed to a constrained dynamical tracking control problem for weight vectors. Instead of non-convex design algorithms, the generalized PID controller structure and improved convex LMI algorithms are proposed to fulfill the PDF tracking problem, meanwhile, in order to enhance robustness, the $L_1$ performance index is used to formulate the disturbance attenuation, which generalizes the corresponding result for linear systems with zero equilibrium (see [133]). Furthermore, a state constraint system related to the characteristics of PDF can also be guaranteed. Rigorous stability and performance analysis are developed via the use of the Lyapunov stability criterion. Simulations are given to demonstrate the efficiency of the proposed approach.

## 3.2 Problem Formulation

### *3.2.1 B-spline Expansion and Dynamic Weight Model*

For a dynamic stochastic system, denote $u(t) \in R^m$ as the control input, $z(t) \in [a,b]$ as the stochastic output and the probability of output $z(t)$ lying inside $[a,y]$ can be described as

$$P(a \le z(t) < y, u(t)) = \int_a^y \gamma(\eta, u(t)) \mathrm{d}\eta \tag{3.1}$$

where $\gamma(y,u(t))$ is the output PDF of the stochastic variable $y(t)$ under control input $u(t)$. As in [63, 64, 157], it is supposed that the output PDF $\gamma(y,u(t))$, as the control objective, can be measured or estimated. For PDF $\gamma(y,u(t))$, the square root B-spline expansion is given by

$$\sqrt{\gamma(y,u(t))} = \sum_{i=1}^{n} \upsilon_i(u(t)) B_i(y) \tag{3.2}$$

where $B_i(y)(i = 1,2,\cdots,n)$ are specified basis functions and $\upsilon_i(t) := \upsilon_i(u(t))(i = 1,2,\cdots,n)$ are the corresponding weights, which depend on $u(t)$. It can be seen that the positiveness of $\gamma(y,u(t))$ can be automatically guaranteed. On the other hand, the PDF should satisfy the condition $\int_a^b \gamma(y,u(t))\,\mathrm{d}y = 1$, which means only $n-1$ weights are independent. So the square root expansions are considered as follows:

$$\gamma(y,u(t)) = (C_0(y)V(t) + v_n(t)B_n(y))^2 \tag{3.3}$$

where $C_0(y) = [B_1(y)\ \ B_2(y)\ \ \cdots\ \ B_{n-1}(y)]$, $V(t) = [v_1(t)\ \ v_2(t)\ \ \cdots\ \ v_{n-1}(t)]^T$.

For simplicity, we denote

$$\Lambda_1 = \int_a^b C_0^T(y)C_0(y)\mathrm{d}y,$$

$$\Lambda_2 = \int_a^b C_0(y)B_n(y)\mathrm{d}y,$$

$$\Lambda_3 = \int_a^b B_n^2(y)\mathrm{d}y \neq 0. \tag{3.4}$$

To guarantee $\int_a^b \gamma(y,u(t))\mathrm{d}y = 1$, $V^T(t)\Lambda_2^T\Lambda_2V(t) - (V^T(t)\Lambda_1V(t) - 1)\Lambda_3 \geq 0$ should be satisfied, which is equivalent to

$$V^T(t)\Pi_0V(t) \leq 1 \tag{3.5}$$

where $\Pi_0 = \Lambda_1 - \Lambda_3^{-1}\Lambda_2^T\Lambda_2 > 0$.

Under condition (3.4), $v_n(t)$ can be represented by a function of $V(t)$ as follows:

$$v_n(t) = h(V(t)) = \frac{\sqrt{\Lambda_3 - V^T(t)\Lambda_0V(t)} - \Lambda_2V(t)}{\Lambda_3} \tag{3.6}$$

where $\Lambda_0 = \Lambda_1\Lambda_3 - \Lambda_2^T\Lambda_2$. It is noted that inequality (3.5) can be considered as a constraint on $V(t)$, which forms one of the key difficulties in the controller design.

The next step is to establish the dynamic model between the control input and the weights. This procedure has been widely used in PDF control and entropy optimization problems. This procedure can be carried out by the corresponding identification processes in [157], or the NN modeling process in [10]. To simply the design algorithm, originally only linear models were considered, where the shape of output PDFs cannot be changed. In this chapter, the following nonlinear dynamic model will be considered as

$$\begin{aligned}\dot{V}(t) &= A_0V(t) + \sum_{i=1}^{N} A_{0d_i}V(t-d_i(t)) + F_0f_0(V(t)) + B_0u(t) \\ &\quad + \sum_{i=1}^{N} B_{0d_i}u(t-d_i(t)) + B_{01}w(t)\end{aligned} \tag{3.7}$$

where $V(t) \in R^{n-1}$ are the independent weight vectors, $u(t)$ and $w(t)$ represent the control input and the exogenous disturbances, respectively. In this chapter, $w(t)$ is supposed to satisfy $\|w(t)\|_\infty = \sup_{t\geq 0}\|w(t)\| < \infty$. $A_0, A_{0d_i}, F_0, B_0, B_{0d_i}$ and $B_{01}$ are known coefficient matrices with compatible dimensions. The time-varying delays $d_i(t)$ satisfy $0 < \dot{d}_i(t) < \beta_i < 1, i = 1,\cdots,N$, we denote $d := \max_{k=1,\cdots,N}\{d_k(0)\}$. $f_0(V(t))$ is a nonlinear function satisfying the following *Lipschitz* condition:

$$\|f_0(V_1(t)) - f_0(V_2(t))\| \leq \|U_0(V_1(t) - V_2(t))\| \tag{3.8}$$

for any $V_1(t)$ and $V_2(t)$, where $U_0$ is a known matrix. It is noted that $f_0(V(t))$ can also be regarded as a kind of unknown modeling uncertainty.

### 3.2.2 Generalized PID Controller Design

A desired PDF $g(y)$ can be given by

$$g(y) = (C_0(y)V_g(t) + h(V_g(t))B_n(y))^2 \tag{3.9}$$

where $V_g(t)$ is the desired weight vector with respect to the same basis function $B_i(y)$. The tracking objective is to find $u(t)$ such that $\gamma(y,u(t))$ can follow $g(y)$. The error between the output PDF and the target PDF is formulated by $\Delta_e(y,t) = \sqrt{\gamma(y,u(t))} - \sqrt{g(y)}$, i.e.,

$$\Delta_e(y,t) = C_0 e(t) + [h(V(t)) - h(V_g(t))]B_n(y) \tag{3.10}$$

where $e(t) = V(t) - V_g(t)$. Due to continuity of $h(V(t))$, $\Delta_e \to 0$ holds as long as $e(t) \to 0$.

For the PDF tracking control problem, classical PID control is unavailable since $\Delta_e(y,t)$ cannot be used as feedback. As a result, we adopt the generalized PID controller

$$u(t) = K_P V(t) + K_I \int_0^t e(\tau)d\tau + K_D \dot{V}(t). \tag{3.11}$$

Based on system (3.7), we introduce a new state variable $x(t) := [\dot{V}^T(t), V^T(t), \int_0^t e^T(\tau)d\tau]^T$. Then the nonlinear weight dynamics can be transformed into an equivalent descriptor form:

$$\begin{cases} E\dot{x}(t) = Ax(t) + \sum_{i=1}^N A_{d_i}x(t-d_i(t)) + Ff(x(t)) + Bu(t) \\ \qquad + \sum_{i=1}^N B_{d_i}u(t-d_i(t)) + B_1 w(t) + HV_g(t) \\ z(t) = Cx(x) + Dw(x) \\ x(t) = \phi(t), \quad t \in [-d, 0] \end{cases} \tag{3.12}$$

where $z(t)$ is the controller output, $\phi(t)$ represents the initial condition of system (3.12), $f(x(t)) = [f_0^T(V(t)), 0, 0]^T$, and

$$E = \begin{bmatrix} 0 & 0 & 0 \\ 0 & I & 0 \\ 0 & 0 & I \end{bmatrix}, \quad A = \begin{bmatrix} -I & A_0 & 0 \\ I & 0 & 0 \\ 0 & I & 0 \end{bmatrix}, \quad A_{d_i} = \begin{bmatrix} 0 & A_{0d_i} & 0 \\ 0 & 0 & 0 \\ 0 & 0 & 0 \end{bmatrix}, \quad F = \begin{bmatrix} F_0 \\ 0 \\ 0 \end{bmatrix},$$

$$B = \begin{bmatrix} B_0 \\ 0 \\ 0 \end{bmatrix}, \quad B_{d_i} = \begin{bmatrix} B_{0d_i} \\ 0 \\ 0 \end{bmatrix}, \quad B_1 = \begin{bmatrix} B_{01} \\ 0 \\ 0 \end{bmatrix}, \quad H = \begin{bmatrix} 0 \\ 0 \\ -I \end{bmatrix}.$$

With such an augmented descriptor system (3.12), the tracking problem can be further reduced to a stabilization control framework because the PID controller can be formulated as

$$u(t) = Kx(t), \quad K = [K_D \;\; K_P \;\; K_I]. \tag{3.13}$$

In the following, we will compute gain matrices $K_P$, $K_I$ and $K_D$ such that the tracking error $e(t)$ converges to zero under state constraints.

*Remark 3.1* It should be pointed out that although some nonlinear tracking approaches have been provided in the last decade [114], less results can be applied to the above model. The key obstacles are that not only the state constraint, but also the modeling uncertainty and the exogenous disturbance have to be considered simultaneously. In the following, we will perform a class of convex LMI algorithms to achieve the above control objectives.

## 3.3 Constrained PID Controller Design Based on LMIs

### 3.3.1 Peak-to-peak Performance Control

$L_1$ performance index is the measure used to describe the level of disturbance attenuation, which is also called peak-to-peak performance (see [133] for details).

**Definition 3.1** The peak-to-peak gain is defined by $\sup_{\|w\|_\infty \le 1} \|z(t)\|_\infty$. The peak-to-peak control problem is to find controller $u(t)$ such that the peak-to-peak gain is minimized or satisfies

$$\sup_{\|w\|_\infty \le 1} \|z(t)\|_\infty < \gamma \tag{3.14}$$

or $\sup_{0 \le \|w\| \le \infty} \frac{\|z(t)\|_\infty}{\|w(t)\|_\infty} < \gamma$, where $\gamma$ is a given positive constant.

Since $V_g(t)$ is a known vector, we denote $y_d := \|V_g(t)\|^2$. It is noted that

$$\|f(x_1(t)) - f(x_2(t))\| \le \|U(x_1(t) - x_2(t))\| \tag{3.15}$$

where $U := \mathrm{diag}\{0, U_0, 0\}$. The following result provides a criterion for the $L_1$ performance problem of the unforced system (3.12).

**Theorem 3.1** For the known parameters $\lambda$, $\mu_i (i = 1,2,3)$, $\alpha > 0$ and matrix $U$, suppose that there exist matrices $P$, $T>0$, $S_i>0$ $i=1,\cdots,N$ and parameter $\gamma>0$ such that the following LMIs

$$P^T E = EP \geq 0, \tag{3.16}$$

$$\begin{bmatrix} \mathrm{sym}(A^T P) + \sum_{i=1}^{N} S_i + \mu_1^2 T & P^T \hat{A}_{d_i} & U^T & P^T F & P^T B_1 & P^T H \\ \hat{A}_{d_i}^T P & \hat{S} & 0 & 0 & 0 & 0 \\ U & 0 & -\lambda^{-2} I & 0 & 0 & 0 \\ F^T P & 0 & 0 & -\lambda^2 I & 0 & 0 \\ B_1^T P & 0 & 0 & 0 & -\mu_2^2 I & 0 \\ H^T P & 0 & 0 & 0 & 0 & -\mu_3^2 I \end{bmatrix} < 0 \tag{3.17}$$

where

$$\hat{A}_{d_i} = [A_{d_1}\ A_{d_2} \cdots A_{d_N}],\quad \hat{S} := \mathrm{diag}\{-(1-\beta_1)S_1, \cdots, -(1-\beta_N)S_N\}$$

and

$$\begin{bmatrix} \mu_1^2 T & 0 & C^T \\ 0 & (\gamma - \mu_2^2 - \mu_3^2 y_d) I & D^T \\ C & D & \gamma I \end{bmatrix} > 0, \tag{3.18}$$

$$\begin{bmatrix} \alpha I & T \\ T & T \end{bmatrix} > 0, \qquad \begin{bmatrix} T & 0 & C^T \\ 0 & (\gamma - \alpha x_m^T x_m) I & D^T \\ C & D & \gamma I \end{bmatrix} > 0 \tag{3.19}$$

are solvable, then the descriptor system is stable, and $\sup_{0\leq\|w\|\leq\infty} \frac{\|z(t)\|_\infty}{\|w(t)\|_\infty} < \gamma$ holds.

*Proof.* Defining a Lyapunov-Krasovskii function as

$$V(x(t),t) = x^T(t) P^T E x(t) + \sum_{i=1}^{N} \int_{t-d_i(t)}^{t} x^T(\tau) S_i x(\tau) \mathrm{d}\tau$$

$$+ \int_0^t [\|\lambda U x(\tau)\|^2 - \|\lambda f(x(\tau))\|^2] \mathrm{d}\tau, \tag{3.20}$$

it is noted that $V(x(t),t) \geq 0$. Furthermore, it can be seen that

$$\frac{\mathrm{d}V(x(t),t)}{\mathrm{d}t} = 2x^T(t) P^T E \dot{x}(t) + \sum_{i=1}^{N} x^T(t) S_i x(t) - \sum_{i=1}^{N} (1 - \dot{d}_i(t)) x_{d_i}^T S_i x_{d_i}$$

$$+ \|\lambda U x(t)\|^2 - \|\lambda f(x(t))\|^2$$

$$= x^T(t)(P^T A + A^T P + \sum_{i=1}^{N} S_i) x(t) + \sum_{i=1}^{N} 2x^T(t) P^T A_{d_i} x_{d_i}(t)$$

$$- \sum_{i=1}^{N} (1 - \dot{d}_i) x_{d_i}^T(t) S_i x_{d_i}(t) + 2x^T(t) P^T F f(x) + 2x^T(t) P^T B_1 w(t)$$

$$+2x^T(t)P^THV_g(t)+\|\lambda Ux(t)\|^2-\|\lambda f(x(t))\|^2$$

$$\leq \zeta^T(t)\Phi_1\zeta(t)+\|\mu_2 w(t)\|^2+\mu_3^2 y_d \tag{3.21}$$

where

$$\zeta(t)=[x^T(t)\ x_{d_1}^T(t)\cdots x_{d_N}^T(t)]^T,\quad \Phi_1=\begin{bmatrix}\Xi_1 & P^T\hat{A}_{d_i}\\ \hat{A}_{d_i}^TP & \hat{S}\end{bmatrix},$$

$$\Xi_1=\mathrm{sym}(A^TP)+\lambda^2U^TU+\frac{1}{\lambda^2}P^TFF^TP+\frac{1}{\mu_2^2}P^TB_1B_1^TP,$$

$$+\frac{1}{\mu_3^2}P^THH^TP+\sum_{i=1}^{N}S_i.$$

Based on Schur complement formula, LMI (3.17) implies that $\Phi_1<\mathrm{diag}[-\mu_1^2T,0]$ holds. With inequality (3.21), it can be seen that for any $w(t)$ satisfying $\|w(t)\|_\infty\leq 1$, we have

$$\frac{\mathrm{d}V(x(t),t)}{\mathrm{d}t}\leq -\mu_1^2x^T(t)Tx(t)+\mu_2^2+\mu_3^2y_d \tag{3.22}$$

where $\mu_3^2y_d$ can be considered as a known parameter. Thus, $\frac{\mathrm{d}V(x(t),t)}{\mathrm{d}t}<0$, if $x^T(t)Tx(t)>\mu_1^{-2}(\mu_2^2+\mu_3^2y_d)$ holds. So we obtain $x^T(t)Tx(t)<\mu_1^{-2}(\mu_2^2+\mu_3^2y_d)$. Therefore for any $x(t)$, it can be verified that

$$x^T(t)Tx(t)\leq \max\{\mathrm{x_m^TTx_m},\mu_1^{-2}(\mu_2^2+\mu_3^2\mathrm{y_d})\},$$

$$\|x_m\|=\sup_{-d\leq t\leq 0}\|x(t)\| \tag{3.23}$$

which also implies that the unforced system (3.12) is stable.

From inequality (3.23), we have that $x^T(t)Tx(t)\leq x_m^TTx_m$ or $x^T(t)Tx(t)\leq \mu_1^{-2}(\mu_2^2+\mu_3^2y_d)$. Combining with LMI (3.18), it can be seen that

$$\begin{bmatrix}\mu_1^2T & 0\\ 0 & (\gamma-\mu_2^2-\mu_3^2y_d)I\end{bmatrix}-\frac{1}{\gamma}\begin{bmatrix}C^T\\ D^T\end{bmatrix}\begin{bmatrix}C & D\end{bmatrix}>0,$$

which guarantees under $x^T(t)Tx(t)\leq \mu_1^{-2}(\mu_2^2+\mu_3^2y_d)$ and $\|w(t)\|_\infty\leq 1$, we have

$$\frac{1}{\gamma}\|z(t)\|^2<\mu_1^2x^T(t)Tx(t)+(\gamma-\mu_2^2-\mu_3^2y_d)w^T(t)w(t)<\gamma. \tag{3.24}$$

On the other hand, from LMI (3.19), it can also be shown that

$$\begin{bmatrix}T & 0\\ 0 & (\gamma-\alpha x_m^Tx_m)I\end{bmatrix}-\frac{1}{\gamma}\begin{bmatrix}C^T\\ D^T\end{bmatrix}\begin{bmatrix}C & D\end{bmatrix}>0.$$

Similarly to the above proof, under $x^T(t)Tx(t)\leq x_m^TTx_m$ and $\|w(t)\|_\infty\leq 1$, we can get

$$\frac{1}{\gamma}\|z(t)\|^2<x^T(t)Tx(t)+(\gamma-\alpha x_m^Tx_m)w^T(t)w(t)<\gamma. \tag{3.25}$$

Hence, the $L_1$ norm of the unforced descriptor system is less $\gamma$. □

### 3.3.2 Peak-to-peak Tracking Performance

Considering the state feedback controller, and substituting $u(t) = Kx(t)$ into system (3.12), the corresponding nonlinear closed-loop descriptor system can be described as

$$\begin{cases} E\dot{x}(t) = (A+BK)x(t) + \sum_{i=1}^{N}(A_{d_i} + B_{d_i}K)x(t-d_i(t)) \\ \qquad\qquad + Ff(x(t)) + B_1 w(t) + HV_g \\ z(t) = Cx(t) + Dw(t) \end{cases} . \tag{3.26}$$

**Theorem 3.2** For the known parameters $\lambda$, $\mu_i (i = 1,2,3)$, $\alpha > 0$ and matrix $U$, suppose that there exist $M = T^{-1} > 0$, $Q = P^{-T}$, $R$, $S_i > 0$ $i = 1,\cdots,N$ and parameter $\gamma > 0$ such that the following LMIs

$$EQ^T = QE \geq 0, \tag{3.27}$$

$$\begin{bmatrix} \mathrm{sym}(AQ^T + BR) + \sum_{i=1}^{N} \mathrm{S}_i & \hat{A}_{d_i}\hat{Q}^T + \hat{B}_{d_i}\hat{R} & QU^T & F & B_1 & H & Q \\ \hat{Q}\hat{A}_{d_i}^T + \hat{R}^T\hat{B}_{d_i}^T & \hat{\mathrm{S}} & 0 & 0 & 0 & 0 & 0 \\ UQ^T & 0 & -\lambda^{-2}I & 0 & 0 & 0 & 0 \\ F^T & 0 & 0 & -\lambda^2 I & 0 & 0 & 0 \\ B_1^T & 0 & 0 & 0 & -\mu_2^2 I & 0 & 0 \\ H^T & 0 & 0 & 0 & 0 & -\mu_3^2 I & 0 \\ Q^T & 0 & 0 & 0 & 0 & 0 & -\mu_1^2 M \end{bmatrix} < 0, \tag{3.28}$$

$$\begin{bmatrix} \mu_1^2 M & 0 & MC^T \\ 0 & (\gamma - \mu_2^2 - \mu_3^2 y_d)I & D^T \\ CM & D & \gamma I \end{bmatrix} > 0, \tag{3.29}$$

$$\begin{bmatrix} \alpha I & I \\ I & M \end{bmatrix} > 0, \quad \begin{bmatrix} M & 0 & MC^T \\ 0 & (\gamma - \alpha x_m^T x_m)I & D^T \\ CM & D & \gamma I \end{bmatrix} > 0 \tag{3.30}$$

are solvable, then the closed-loop descriptor system (2.26) is stable and satisfies both $\lim_{t\to\infty} V(t) = V_g(t)$ and $\sup_{0 \leq \|w\| \leq \infty} \frac{\|z(t)\|_\infty}{\|w(t)\|_\infty} < \gamma$. In this case, the PID control gain $K$ can be solved via $R = KQ^T$, and $\mathrm{S}_i = QS_iQ^T$, $\hat{\mathrm{S}} = Q\hat{S}Q^T$, $\hat{Q} = \mathrm{diag}\{Q,\cdots,Q\}$, $\hat{R} = \mathrm{diag}\{R,\cdots,R\}$.

*Proof.* Based on Theorem 3.1 and Lyapunov-Krasovskii function (3.20), we obtain

$$\frac{dV(x(t),t)}{dt} \leq \zeta^T(t)\Phi_2\zeta(t) + \|\mu_2 w(t)\|^2 + \mu_3^2 y_d \tag{3.31}$$

where $\hat{A}_{d_i} = [A_{d_1},\cdots,A_{d_N}]$, $\hat{B}_{d_i} = [B_{d_1},\cdots,B_{d_N}]$, $\hat{K} = \mathrm{diag}\{K,K,\cdots,K\}$ and

$$\Phi_2 = \begin{bmatrix} \Xi_1 + P^T BK + K^T B^T P & P^T \hat{A}_{d_i} + P^T \hat{B}_{d_i} \hat{K} \\ \hat{A}_{d_i}^T P + \hat{K}^T \hat{B}_{d_i}^T P & \hat{S} \end{bmatrix}.$$

By pre-multiplying by diag$\{P^T,\cdots,P^T,I,I,I,I,I\}$ and post-multiplying by diag$\{P,\cdots,P,I,I,I,I,I\}$ on both sides of LMI (3.28), and based on Schur complement formula, it can be seen that for any $w(t)$ satisfying $\|w(t)\|_\infty \leq 1$, we have

$$\frac{\mathrm{d}V(x(t),t)}{\mathrm{d}t} \leq -\mu_1^2 x^T(t)Tx(t) + \mu_2^2 + \mu_3^2 y_d. \tag{3.32}$$

Similarly to the proof of Theorem 3.1, it can be seen that inequality (3.23) still holds for the closed-loop descriptor system, which implies that system (3.26) is still stable in the presence of $w(t)$ and $V_g(t)$. Meanwhile, by pre-multiplying by diag$\{T,I,I\}$, diag$\{I,T\}$ and post-multiplying by diag$\{T,I,I\}$, diag$\{I,T\}$ on both sides of LMIs (3.29) and (3.30), it is easy to show that LMIs (3.29) and (3.30) are equivalent to LMIs (3.18) and (3.19) so that the system (3.26) satisfies the peak-to-peak disturbance attenuation performance.

For a couple of $w(t)$ and $V_g(t)$, we suppose that $\theta_1(t)$ and $\theta_2(t)$ are two trajectories of the closed-loop system corresponding to a fixed initial condition. Defining $\sigma(t) := \theta_1(t) - \theta_2(t)$, then the dynamics for $\sigma(t)$ can be described as

$$E\dot{\sigma}(t) = (A+BK)\sigma(t) + \sum_{i=1}^{N}(A_{d_i} + B_{d_i}K)\sigma(t-d_i(t))$$
$$+F(f(\theta_1(t)) - f(\theta_2(t))). \tag{3.33}$$

Similarly to Equation 3.20, a Lyapunov function can be constructed as

$$V(\sigma(t),t) = \sigma^T(t)P^T E\sigma(t) + \sum_{i=1}^{N}\int_{t-d_i(t)}^{t} \sigma^T(\tau)S_i\sigma(\tau)\mathrm{d}\tau$$
$$+\int_0^t [\|\lambda U\sigma(\tau)\|^2 - \|\lambda(f(\theta_1(\tau)) - f(\theta_2(\tau)))\|^2]\mathrm{d}\tau. \tag{3.34}$$

Based on LMI (3.28) and similarly to the above proof, we can get

$$\frac{\mathrm{d}V(\sigma(t),t)}{\mathrm{d}t} \leq -\mu_1^2 \sigma^T(t)T\sigma(t) \leq -\mu_1^2 \lambda_{\min}(T)\|\sigma(t)\|^2 \tag{3.35}$$

where $\lambda_{\min}(T)$ denotes the minimum eigenvalue of $T$. It can be verified that $\sigma = 0$ is the unique asymptotically stable equilibrium point of system (3.33). It means that the closed-loop system (3.26) also has a unique asymptotically stable equilibrium point. Furthermore, it can be concluded that $\lim_{t\to\infty}\frac{\mathrm{d}}{\mathrm{d}t}(\int_0^t e(\tau)\mathrm{d}\tau) = 0$, which shows that $\lim_{t\to\infty} V(t) = V_g(t)$. □

### 3.3.3 Peak-to-peak Constrained Tracking Control

It has been shown that due to the property of the PDF, the weight vectors have to satisfy $V^T(t)\Pi_0 V(t) \leq 1$, which can be reduced to $x^T(t)\Pi x(t) \leq 1$, where $\Pi := \mathrm{diag}\{0, \Pi_0, 0\}$. Based on the property of non-negative definite matrix, $\Pi$ can be divided into $\Pi = G^2$, where $G \geq 0$. For the nonlinear closed-loop system described by Equation 3.26, the following result provides an algorithm to design the tracking controller with guaranteed state constraints and disturbance attenuation performance.

**Theorem 3.3** For the known parameters $\lambda$, $\mu_i (i = 1,2,3)$, $\alpha > 0$ and matrix $U$, $G$, suppose that there exist $M = T^{-1} > 0$, $Q = P^{-T}$, $R$, $S_i > 0$ $i = 1, \cdots, N$ and parameter $\gamma > 0$ such that (3.27)-(3.30) and the following LMIs

$$\begin{bmatrix} M & MG \\ GM & \mu_1^2(\mu_2^2 + \mu_3^2 y_d)^{-1} I \end{bmatrix} \geq 0, \tag{3.36}$$

$$\begin{bmatrix} M & MG \\ GM & I \end{bmatrix} \geq 0, \quad \begin{bmatrix} 1 & x_m^T \\ x_m & M \end{bmatrix} \geq 0 \tag{3.37}$$

are solvable, then closed-loop system (3.26) is stable, satisfies $x^T(t)\Pi x(t) \leq 1$, $\lim_{t\to\infty} V(t) = V_g$ and $\sup_{0 \leq \|w\| \leq \infty} \frac{\|z(t)\|_\infty}{\|w(t)\|_\infty} < \gamma$. In this case, the PID control gain $K$ can be solved via $R = KQ^T$.

*Proof.* It can be shown that the stability, tracking performance and robustness for the closed-loop have been guaranteed. It remains to show that under conditions (3.36) and (3.37), the closed-loop system satisfies the desired constraint.

Similarly to the above proof, inequality (3.23) still holds. This means that $x^T(t)Tx(t) \leq x_m^T T x_m$ or $x^T(t)Tx(t) \leq \mu_1^{-2}(\mu_2^2 + \mu_3^2 y_d)$. Combining with LMI (3.37), $\Pi \leq T$ and $x_m^T T x_m \leq 1$ can be satisfied, furthermore, we can get $x^T(t)\Pi x(t) \leq x^T(t)Tx(t) \leq x_m^T T x_m \leq 1$. On the other hand, from LMI (3.36), we can get $\Pi \leq \mu_1^2(\mu_2^2 + \mu_3^2 y_d)^{-1} T$, and it can be seen that $x^T(t)\Pi x(t) \leq \mu_1^2(\mu_2^2 + \mu_3^2 y_d)^{-1} x^T(t)Tx(t)$ $\leq 1$ can be satisfied. □

*Remark 3.2* Using the LMI-toolbox in MATLAB, the design procedures for the PID controller are given as follows:

1. Solve LMIs (3.27-3.30) and (3.36, 3.37) for $M > 0$, $Q$, $R$, $S_i > 0$.
2. The PID control gain $K$ is solved via $R = KQ^T$ and the control input is provided by Equation 3.11.

The above result shows that the design procedures can be reduced to a class of LMI algorithms with respect to $Q$ and $R$, which is more beneficial than the previous results in [63, 64, 164]. For example, the algorithms provided in [63] were required to solve a class of nonlinear matrix inequalities (NLMIs), which led to a non-convex computation procedure. On the other hand, the proposed results have independent significance for the constrained PI/PID tracking context of the deterministic weighting models, and have potential applications in other PI/PID type control problems.

## 3.4 Simulations

Suppose that the output PDFs can be approximated using the square root B-spline models described by Equation 3.3 with $n = 3$, $y \in [0, 1.5]$, $i = 1, 2, 3$

$$B_i(y) = \begin{cases} |\sin 2\pi y| & y \in [0.5(i-1); 0.5i] \\ 0 & y \in [0.5(j-1); 0.5j] \quad i \neq j. \end{cases}$$

From the notation in condition (3.4), it can be seen that $\Lambda_1 = \mathrm{diag}\{0.25, 0.25\}$, $\Lambda_2 = [0, 0]$, $\Lambda_3 = 0.25$. The desired PDF $g(y)$ is supposed to be described by Equation 3.10 with $V_g = [0.8, 1]^T$.

The dynamic nonlinear model relating $u(t)$ and $V(t)$ is described by condition (3.8) with

$$A_0 = [-0.5\ 0.5, 0.5\ 1.5],\ \ A_{0d_1} = \mathrm{diag}\{-0.5, -0.5\},$$

$$F_0 = \mathrm{diag}\{-0.3, -0.3\},\ \ B_0 = \mathrm{diag}\{0.5, 0.5\},\ \ B_{0d_1} = \mathrm{diag}\{-0.75, -0.75\},$$

$$B_1 = \mathrm{diag}\{-0.5, -0.7\},\ \ U_0 = \mathrm{diag}\{0.3, 0.5\}.$$

Defining $\lambda_1 = 2$, $\mu_1^2 = \mu_2^2 = \mu_3^2 = 2$, $\gamma = 6$ and solving the LMIs (3.28-3.31, 3.37, 3.38), we obtain

$$K_P = \begin{bmatrix} -0.4781 & 1.7281 \\ 0.3550 & -2.8292 \end{bmatrix}, K_I = \begin{bmatrix} -6.3294 & 1.7274 \\ 0.2760 & -5.7269 \end{bmatrix},$$

$$K_D = \begin{bmatrix} 1.0038 & -0.1030 \\ -0.0249 & 1.0202 \end{bmatrix}.$$

When the robust PID controller is applied, the responses of dynamical weight vectors are shown in Figure 3.1 where the top curve shows the response of $v_1$ that converges to 1.0. The bottom curve displays the response of $v_2$, which approaches 0.8 as expected. To achieve the responses shown in Figure 3.1, the feedback control input has the response shown in Figure 3.2 and such responses are generated using the PID control strategy described by Equation 3.13. Under the proposed robust control strategy, the 3D mesh plot of practical PDF is shown in Figure 3.3, where the dynamic performance of the output PDF is given, which shows that the actual output PDF can approach its target PDF accurately. It is demonstrated that satisfactory tracking performance, stability and robustness are achieved. Of course, these responses are oscillatory which shows that the closed-loop system is underdamped. This is because the selected control parameters are obtained as a result of solution of the LMI problem. In practice, fine tuning of these control parameters can be done to improve the closed-loop response.

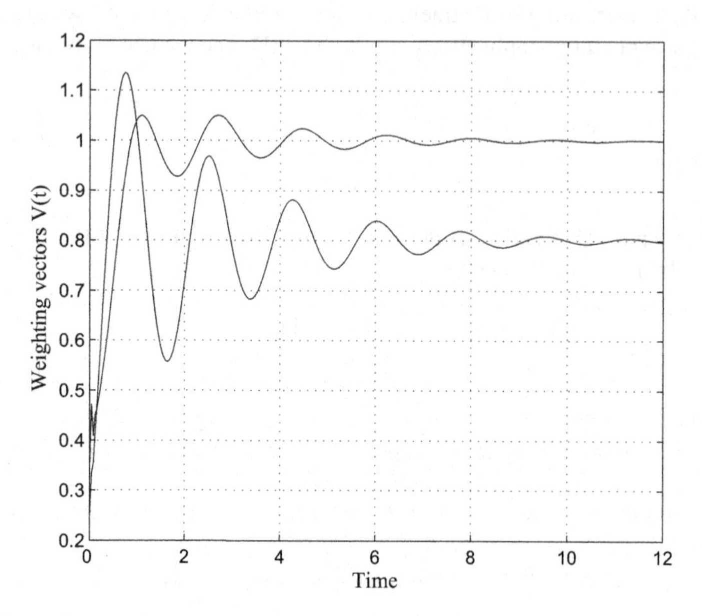


**Figure 3.1** Responses of the weight vectors

## 3.5 Conclusions

In this chapter, the SDC is considered for general non-Gaussian stochastic systems, where the PDF tracking is transformed into a tracking problem with a constraint of partial weight vectors. Compared with previous work, the main results here have four features:

1. New realizable generalized PID controller structures are proposed for PDF tracking control.
2. Using improved convex LMI algorithms, multiple control objectives including stabilization, tracking performances, robustness and state constraints can be guaranteed simultaneously.
3. Exogenous disturbances, state constraints and non-zero equilibrium are all considered in a nonlinear weighting model.
4. To enhance robust performance, the peak-to-peak measure is applied to optimize tracking performance.

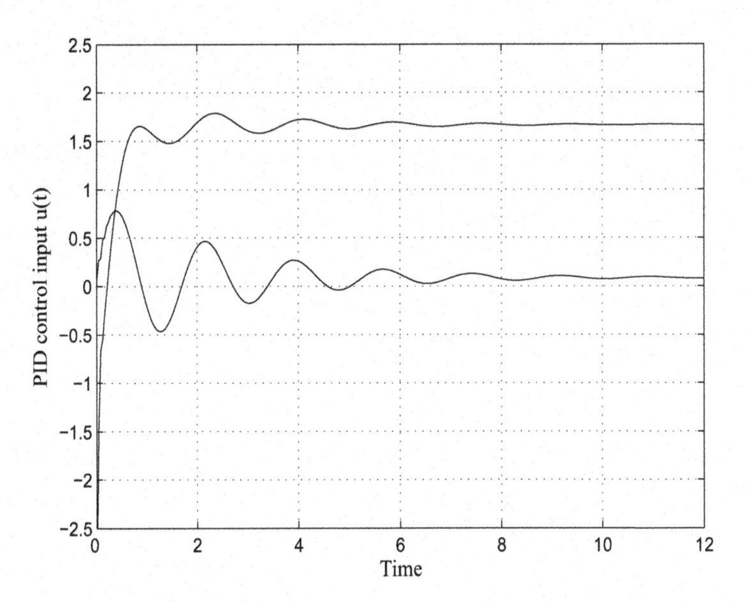


**Figure 3.2** PID-type control input

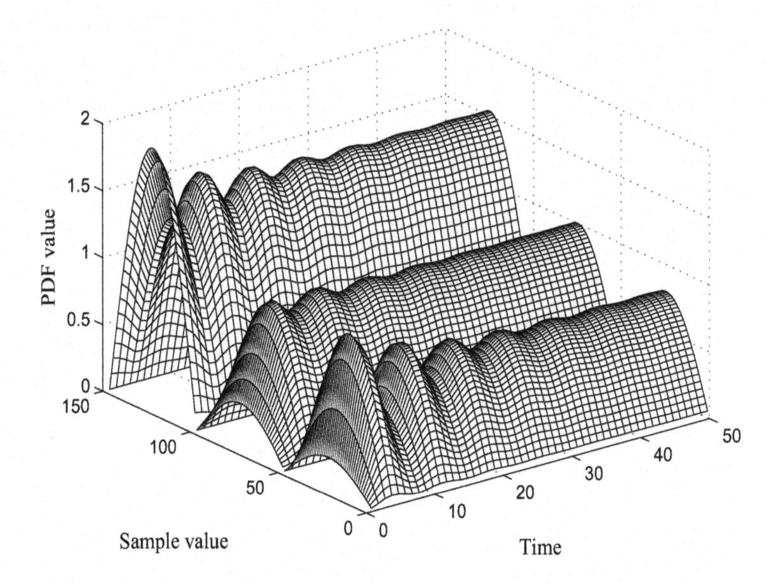


**Figure 3.3** 3D mesh plot of the PDFs

# Chapter 4
# Constrained Discrete-time Proportional Integral Control Based on Convex Algorithms

## 4.1 Introduction

Following the recent developments on PDF shape control, new realizable generalized PI control structures are proposed for SDC problems in the discrete-time context. Here we use the square root B-spline expansion and nonlinear discrete-time weight system to identify the unknown dynamical relationship between the control input and the output PDF. As a result, the tracking control strategy is studied for the complex discrete-time weight models where there exist a time delay term, non-zero equilibrium, partial state constraint, exogenous disturbance and model uncertainty. Improved LMI algorithms are used to design a generalized discrete-time PI controller such that stabilization, tracking performance, robustness and state constraints can be guaranteed simultaneously. Furthermore, the $L_1$ performance index is used to formulate the disturbance attenuation, which generalizes the corresponding result for linear systems with zero equilibrium (see [133]). It is noted that the proposed result represents a significant extension of the previous results [61, 63, 64], and has also independent significance in robust discrete-time PI tracking control fields.

## 4.2 Problem Formulation

### *4.2.1 B-spline Expansion and Discrete-time Weight Dynamical Model*

As defined in [64], let the uniformly bounded random process variable $z(k)$ represent the output of a stochastic system and $u(k) \in R^m$ stand for the input vector that controls the shape of the PDF of $z(k), (k = 0, 1, 2 \cdots)$. At each sample time $k$, output $\eta(k)$ can be characterized by its PDF $\gamma(y, u(k))$, which is assumed to be measurable and defined on a known interval $[a, b]$. Among the proposed B-spline expansions (see [64]), the square root model is more beneficial than the others in terms of positiveness and robustness, where the output PDF $\gamma(y,u(k))$ can be expanded as

$$\sqrt{\gamma(y,u(k))} = \sum_{i=1}^{n} \upsilon_i(u(k))B_i(y) \tag{4.1}$$

where $B_i(y)(i=1,2,\cdots,n)$ are pre-specified basis functions, $y \in [a,b]$ and $\upsilon_i(k) := \upsilon_i(u(k)), (i=1,\cdots,n)$ are the corresponding weight vectors. Because $\int_a^b \gamma(y,u(k))dy = 1$, only $n-1$ weights are independent. Equation 4.1 can be rewritten as

$$\gamma(y,u(k)) = (C_0(y)V(k) + \upsilon_n(k)B_n(y))^2 \tag{4.2}$$

where

$$C_0(y) = [B_1(y)\ B_2(y)\ \cdots\ B_{n-1}(y)],$$
$$V(k) = [\upsilon_1(k)\ \upsilon_2(k)\ \cdots\ \upsilon_{n-1}(k)]^T.$$

Denote

$$\Lambda_1 = \int_a^b C_0^T(y)C_0(y)dy, \quad \Lambda_2 = \int_a^b C_0(y)B_n(y)dy, \quad \Lambda_3 = \int_a^b B_n^2(y)dy. \tag{4.3}$$

To guarantee $\int_a^b \gamma(y,u(k))dy = 1$, $V^T(k)\Lambda_2\Lambda_2^T V(k) - (V^T(k)\Lambda_1 V(k) - 1)\Lambda_3 \geq 0$ should be satisfied, which is equivalent to

$$V^T(k)\Pi_0 V(k) \leq 1, \quad \Pi_0 = \Lambda_1 - \Lambda_3^{-1}\Lambda_2^T\Lambda_2 > 0. \tag{4.4}$$

It is noted that the condition $\Pi_0 > 0$ can be guaranteed by selecting the basis functions recursively and changing the order of the basis functions. Under condition (4.4), $\upsilon_n(t)$ can be represented by a function of $V(k)$ (see [64, 65]):

$$\upsilon_n(k) = h(V(k)) = \frac{\sqrt{\Lambda_3 - V^T(k)\Lambda_0 V(k)} - \Lambda_2 V(k)}{\Lambda_3} \tag{4.5}$$

where $\Lambda_0 = \Lambda_1\Lambda_3 - \Lambda_2^T\Lambda_2$. It is noted that inequality (4.4) can be considered as a constraint on $V(t)$, which constitutes one of the major difficulties in the controller design.

Once B-spline expansions have been made for the output PDFs, the next step is to find the dynamic relationships between the control input and the weight vectors related to the PDFs, which corresponds to a further modeling procedure. However, most published results only concern linear models for the weight dynamics, while generally speaking, the mapping from the control input to the weights may be nonlinear and such a relationship may contain some uncertainties. As such, the nonlinear dynamic model that links the weight vector $V(k)$ with the control input $u(k)$ will be considered as

$$V(k+1) = A_0 V(k) + \sum_{i=1}^{N} A_{0d_i} V(k-d_i) + F_0 f_0(V(k))$$

$$+B_0 u(k) + \sum_{i=1}^{N} B_{0d_i} u(k - d_i) + E_0 w(k) \tag{4.6}$$

where $V(k) \in R^{n-1}$ is the independent weight vector, $u(k)$ and $w(k)$ represent the control input and the exogenous disturbances respectively. In the chapter, $w(k)$ is assumed to satisfy $\|w\|_\infty = \sup_{k \geq 0} \|w(k)\| < \infty$. $A_0, A_{0d_i}, F_0, B_0, B_{0d_i}$ and $E_0$ are known coefficient matrices with compatible dimensions, the time delays $d_i$ satisfy $d := \max_i \{d_i\}$. $f_0(V(k))$ is an unknown nonlinear function satisfying globally the Lipschitz condition

$$\|f_0(V_1(k)) - f_0(V_2(k))\| \leq \|U_0(V_1(k) - V_2(k))\| \tag{4.7}$$

for any $V_1(k)$, $V_2(k)$ and $f_0(0) = 0$, where $U_0$ is a known matrix. It is noted that $f_0(V(k))$ can also be regarded as a kind of unknown modeling uncertainty.

*Remark 4.1* The PDF $\gamma(y, u(k))$, as the control objective, can be measured or estimated by using instruments (such as a laser particle size distribution measure or a digital camera) or the kernel estimation technique based on an open loop test (see [64] for details). Meanwhile, the advantages of the shape control of output PDFs have also been recognized in that the PDF shape includes complete information about both the system dynamics and the system distributions. Traditional statistics such as mean and variance are easily calculated from the PDF, while the PDF shape comprises a wealth of additional information.

### 4.2.2 Target Weight Model and Discrete-time PI Controller

Corresponding to Equation 4.2, the desired PDF $g(y)$ is given by

$$g(y) = (C_0(y)V_g + h(V_g)B_n(y))^2 \tag{4.8}$$

where $V_g$ is the desired weight vector corresponding to $B_i(y)$. The tracking objective is to find $u(k)$ at sample time $k$ such that $\gamma(y, u(k))$ can follow $g(y)$. The error between the output PDFs and the target one can be formulated as

$$e(y,k) = \sqrt{g(y)} - \sqrt{\gamma(y, u(k))} = C_0(y)W_e(k) + [h(V_g) - h(V(k))]B_n(y) \tag{4.9}$$

where $W_e(k) = V_g - V(k)$. Due to continuity of $h(V(k))$, $e(y,k) \to 0$ holds as long as $W_e(k) \to 0$. Thus, the problem will be transformed into a tracking control problem for weight vectors and the objective is to find $u(k)$ such that $e(y,k) \to 0$, when $k \to +\infty$ or $W_e(k) \to 0$, when $k \to +\infty$.

For the PDF tracking control problem, the classical PI control law is unrealizable since information on the error $e(y,k)$ and $e(y, k-1)$ cannot be used as the feedback signal. As a result, the following generalized PI control structure for the weight dynamical systems will be adopted:

$$\begin{cases} u(k) = K_P V(k) + K_I \xi(k) \\ \xi(k) = \xi(k-1) + T_0 W_e(k-1) \end{cases} \tag{4.10}$$

where $T_0 = t_0 I$, with $t_0 > 0$ being the sample interval, and $\xi(k)$ corresponds to the integral term in [198]. At this stage, the PDF tracking control can be reduced to the constrained PI tracking control problem for nonlinear discrete-time systems (4.6), and the objective is to find gain matrices $K_P$, $K_I$ such that the nonlinear closed-loop system is stable and the tracking error $W_e(k)$ converges to zero under state constraints.

Based on system (4.6), we introduce a new state variable $x(k) := [V^T(k), \xi^T(k)]^T$. Then the nonlinear weight system can be described by

$$\begin{cases} x(k+1) = Ax(k) + \sum_{i=1}^{N} A_{d_i} x(k-d_i) + Ff(x(k)) + Bu(k) \\ \qquad\qquad + \sum_{i=1}^{N} B_{d_i} u(k-d_i) + Ew(k) + HV_g \\ z(k) = Cx(k) + Dw(k) \\ x(k) = \varphi(k), \quad k \in [-d, 0] \end{cases} \tag{4.11}$$

where $z(k)$ is the control output, and

$$A = \begin{bmatrix} A_0 & 0 \\ -T_0 & I \end{bmatrix}, \quad A_{d_i} = \begin{bmatrix} A_{0d_i} & 0 \\ 0 & 0 \end{bmatrix}, \quad B = \begin{bmatrix} B_0 \\ 0 \end{bmatrix}, \quad B_{d_i} = \begin{bmatrix} B_{0d_i} \\ 0 \end{bmatrix},$$

$$F = \begin{bmatrix} F_0 \\ 0 \end{bmatrix}, \quad E = \begin{bmatrix} E_0 \\ 0 \end{bmatrix}, \quad H = \begin{bmatrix} 0 \\ T_0 \end{bmatrix}.$$

*Remark 4.2* It should be pointed out that although nonlinear tracking control theory in the classical context (rather than in PDF control) has been widely developed, less results can be applied to the above nonlinear discrete-time model. The key obstacles are that not only the state constraint, but also the exogenous disturbance have to be considered simultaneously. On the other hand, time delays are frequently encountered in many practical engineering systems. It is now well known that time delay is one of the main causes of instability and poor performance of control systems. In our opinion, insertion of the time delay term into weight system (4.6) is reasonable and significant.

## 4.3 Robust Constrained Tracking Control

### *4.3.1 Solvability for Peak-to-peak Performance*

**Definition 4.1** The peak-to-peak gain is defined by $\sup_{\|w\|_\infty \le 1} \|z(k)\|_\infty$. The peak-to-peak control problem is to find controller $u(k) = Kx(k)$ at sample time $k$ such that the peak-to-peak gain is minimized or satisfies $\sup_{\|w\|_\infty \le 1} \|z(k)\|_\infty \le \gamma$ or $\sup_{0 \le \|w\| \le \infty} \frac{\|z(k)\|_\infty}{\|w(k)\|_\infty} \le \gamma$, where $\gamma > 0$ is a given constant.

Since $V_g$ can be seen as a known vector, we denote $y_d := \|V_g\|^2$. Based on Equation 4.7, we can get

$$\|f(x_1(k)) - f(x_2(k))\| \leq \|U(x_1(k) - x_2(k))\| \tag{4.12}$$

for any $x_1(k)$, $x_2(k)$ and $f(0) = 0$, where $U := \text{diag}\{U_0, 0\}$. The following result provides a criterion for the $L_1$ performance problem of the unforced system (4.11), which also generalizes the corresponding result for linear systems with zero equilibrium and in the absence of state constraints.

**Theorem 4.1** For the known parameters $\lambda$, $\mu_i (i = 1,2,3)$, $\alpha > 0$ and matrix $U$, suppose that there exist matrices $T > 0$, $P > 0$, $S_i > 0$ $i = 1, \cdots, N$ and parameter $\gamma > 0$, such that the following LMIs

$$\begin{bmatrix} -P + \sum_{i=1}^{N} S_i + \mu_3^2 T + \lambda^2 U^T U & 0 & 0 & 0 & 0 & A^T P \\ 0 & -\hat{S} & 0 & 0 & 0 & \hat{A}_d^T P \\ 0 & 0 & -\lambda^2 I & 0 & 0 & F^T P \\ 0 & 0 & 0 & -\mu_1^2 I & 0 & E^T P \\ 0 & 0 & 0 & 0 & -\mu_2^2 I & H^T P \\ PA & P\hat{A}_d & PF & PE & PH & -P \end{bmatrix} < 0, \tag{4.13}$$

and

$$\begin{bmatrix} \mu_3^2 T & 0 & C^T \\ 0 & (\gamma - \mu_1^2 - \mu_2^2 y_d) I & D^T \\ C & D & \gamma I \end{bmatrix} > 0, \tag{4.14}$$

and

$$\begin{bmatrix} \alpha I & T \\ T & T \end{bmatrix} > 0, \quad \begin{bmatrix} T & 0 & C^T \\ 0 & (\gamma - \alpha x_m^T x_m) I & D^T \\ C & D & \gamma I \end{bmatrix} > 0 \tag{4.15}$$

are solvable, then the unforced nonlinear system (4.11) is stable, and satisfies $\sup_{0 \leq \|w\| \leq \infty} \frac{\|z(k)\|_\infty}{\|w(k)\|_\infty} < \gamma$.

*Proof.* Select a Lyapunov-Krasovskii function as

$$V(x(k),k) = x^T(k) P x(k) + \sum_{i=1}^{N} \sum_{l=1}^{d_i} x^T(k-l) S_i x(k-l)$$

$$+ \sum_{i=1}^{k-1} [\|\lambda U x(i)\|^2 - \|\lambda f(x(i))\|^2]. \tag{4.16}$$

Obviously both the first term and the second term of Equation 4.16 are non-negative. Based on the inequality (4.12), we can get $\|f(x(k))\| \leq \|U(x(k)\|$. So it can be verified that $V(x(k),k) \geq 0$. Furthermore, it can be seen that

$$\Delta V(x(k),k) = V(x(k+1),k+1) - V(x(k),k)$$

$$
\begin{aligned}
&= x^T(k+1)Px(k+1) - x^T(k)Px(k) + \|\lambda Ux(k)\|^2 - \|\lambda f(x(k))\|^2 \\
&\quad + \sum_{i=1}^{N}[\sum_{l=0}^{d_i-1} x^T(k-l)S_ix(k-l) - \sum_{l=1}^{d_i} x^T(k-l)S_ix(k-l)] \\
&= x^T(k+1)Px(k+1) - \sum_{i=1}^{N} x^T(k-d_i)S_ix(k-d_i) \\
&\quad + x^T(k)(\sum_{i=1}^{N} S_i - P)x(k) + \|\lambda Ux(k)\|^2 - \|\lambda f(x(k))\|^2 \\
&= \zeta^T(k)\Upsilon_1\zeta(k) + \mu_1^2\|w(k)\|^2 + \mu_2^2\|V_g\|^2
\end{aligned}
\tag{4.17}
$$

where

$$
\zeta(k) = [x^T(k)\ \hat{x}_d^T(k)\ f^T(x(k))\ w^T(k)\ V_g^T]^T,
$$

$$
\hat{x}_d^T(k) = [x^T(k-d_1), \cdots, x^T(k-d_N)],
$$

$$
\hat{A}_d = [A_{d_1}\ A_{d_2} \cdots, A_{d_N}],\quad \hat{S} := diag\{S_1,\ S_1, \cdots, S_N\},
$$

$$
\Upsilon_1 = \begin{bmatrix}
A^TPA + \sum_{i=1}^{N} S_i - P + \lambda^2U^TU & A^TP\hat{A}_d & A^TPF & A^TPE & A^TPH \\
\hat{A}_d^TPA & \hat{A}_d^TP\hat{A}_d - \hat{S} & \hat{A}_d^TPF & \hat{A}_d^TPE & \hat{A}_d^TPH \\
F^TPA & F^TP\hat{A}_d & F^TPF - \lambda^2I & F^TPE & F^TPH \\
E^TPA & E^TP\hat{A}_d & E^TPF & E^TPE - \mu_1^2I & E^TPH \\
H^TPA & H^TP\hat{A}_d & H^TPF & H^TPE & H^TPH - \mu_2^2I
\end{bmatrix}.
$$

Using the schur complement, we have $\Upsilon_1 < \mathrm{diag}[-\mu_3^2T, 0, 0, 0, 0] \Longleftrightarrow (4.13)$. With Equation 4.17, it can be seen that for any $w(k)$ satisfying $\|w(k)\|_\infty \leq 1$, we have

$$
\Delta V(x(k), k) \leq -\mu_3^2x^T(k)Tx(k) + \mu_1^2 + \mu_2^2y_d. \tag{4.18}
$$

Thus, $\Delta V(x(k), k) < 0$, if $x^T(k)Tx(k) > \mu_3^{-2}(\mu_1^2 + \mu_2^2y_d)$ holds. So for any $x(k)$, it can be verified that

$$
x^T(k)Tx(k) \leq \max\{x_m^TTx_m, \mu_3^{-2}(\mu_1^2 + \mu_2^2y_d)\},
$$

$$
\|x_m\| = \sup_{-d \leq k \leq 0}\|x(k)\|, \tag{4.19}
$$

which also implies that the unforced system (4.11) is stable.

From inequality (4.19), we obtain $x^T(k)Tx(k) \leq x_m^TTx_m$ or $x^T(k)Tx(k) \leq \mu_3^{-2}(\mu_1^2 + \mu_2^2y_d)$. Combining with LMI (4.14), it can be seen that

$$
\begin{bmatrix} \mu_3^2T & 0 \\ 0 & (\gamma - \mu_1^2 - \mu_2^2y_d)I \end{bmatrix} - \frac{1}{\gamma}\begin{bmatrix} C^T \\ D^T \end{bmatrix}[C\ D] > 0,
$$

which guarantees under $x^T(k)Tx(k) \leq \mu_3^{-2}(\mu_1^2 + \mu_2^2y_d)$ and $\|w(k)\|_\infty \leq 1$, we have

$$
\frac{1}{\gamma}\|z(k)\|^2 < \mu_3^2x^T(k)Tx(k) + (\gamma - \mu_1^2 - \mu_2^2y_d)w^T(k)w(k) \leq \gamma. \tag{4.20}
$$

On the other hand, from LMI (4.15), it can also be shown that

$$\begin{bmatrix} T & 0 \\ 0 & (\gamma - \alpha x_m^T x_m)I \end{bmatrix} - \frac{1}{\gamma} \begin{bmatrix} C^T \\ D^T \end{bmatrix} \begin{bmatrix} C & D \end{bmatrix} > 0.$$

Similarly to the above proof, under $x^T(k)Tx(k) \le x_m^T T x_m$ and $\|w(k)\|_\infty \le 1$, we obtain

$$\begin{aligned} \frac{1}{\gamma}\|z(t)\|^2 &< x^T(k)Tx(k) + (\gamma - \alpha x_m^T x_m) w^T(k) w(k) \\ &\le \alpha x_m^T x_m + (\gamma - \alpha x_m^T x_m) w^T(k) w(k) = \gamma. \end{aligned} \tag{4.21}$$

Hence, the peak-to-peak gain of the unforced system is less than $\gamma$. □

### 4.3.2 Peak-to-peak Tracking Performance

Considering the state feedback controller with PI control structure, and substituting $u(k) = Kx(k)$ into system (4.11), the corresponding nonlinear closed-loop system can be described by

$$\begin{cases} x(k+1) = (A+BK)x(k) + \sum_{i=1}^{N}(A_{d_i} + B_{d_i}K)x(k-d_i) \\ \qquad\qquad +Ff(x(k)) + Ew(k) + HV_g \\ z(k) = Cx(k) + Dw(k) \end{cases} . \tag{4.22}$$

The following result provides a solution for the nonlinear tracking control problem with disturbance attenuation performance.

**Theorem 4.2** For the known parameters $\lambda$, $\mu_i (i = 1,2,3)$, $\alpha > 0$ and matrix $U$, suppose that there exist $M = T^{-1} > 0$, $Q = P^{-1} > 0$, $R$, $S_i > 0$ $i = 1,\cdots,N$ and parameter $\gamma > 0$ such that the following LMIs

$$\begin{bmatrix} -Q + \sum_{i=1}^{N} QS_iQ & 0 & 0 & 0 & 0 & QA^T + RB^T & QU^T & Q \\ 0 & -\hat{Q}\hat{S}\hat{Q} & 0 & 0 & 0 & \hat{Q}\hat{A}_d^T + \hat{R}\hat{B}_d^T & 0 & 0 \\ 0 & 0 & -\lambda^2 I & 0 & 0 & F^T & 0 & 0 \\ 0 & 0 & 0 & -\mu_1^2 I & 0 & E^T & 0 & 0 \\ 0 & 0 & 0 & 0 & -\mu_2^2 I & H^T & 0 & 0 \\ AQ + BR^T & \hat{A}_d\hat{Q} + \hat{B}_d\hat{R}^T & F & E & H & -Q & 0 & 0 \\ UQ & 0 & 0 & 0 & 0 & 0 & -\lambda^{-2} I & 0 \\ Q & 0 & 0 & 0 & 0 & 0 & 0 & -\mu_3^{-2} M \end{bmatrix} < 0, \tag{4.23}$$

and

$$\begin{bmatrix} \mu_3^2 M & 0 & MC^T \\ 0 & (\gamma - \mu_1^2 - \mu_2^2 y_d)I & D^T \\ CM & D & \gamma I \end{bmatrix} > 0, \tag{4.24}$$

and

$$\begin{bmatrix} \alpha I & I \\ I & M \end{bmatrix} > 0, \quad \begin{bmatrix} M & 0 & MC^T \\ 0 & (\gamma - \alpha x_m^T x_m)I & D^T \\ CM & D & \gamma I \end{bmatrix} > 0 \tag{4.25}$$

are solvable, then the closed-loop system (4.22) is stable and satisfies both $\lim_{k\to\infty} V(k) = V_g$ and $\sup_{0\leq\|w\|\leq\infty} \frac{\|z(k)\|_\infty}{\|w(k)\|_\infty} < \gamma$. In this case, the PI control gain $K$ can be solved via $R = QK^T$, and $\hat{Q} = \mathrm{diag}\{Q,\cdots,Q\}$, $\hat{R} = \mathrm{diag}\{R,\cdots,R\}$.

*Proof.* Based on Theorem 4.1 and Lyapunov-Krasovskii function (4.16), we can get

$$\Delta V(x(k),k) \leq \zeta^T(k)\Upsilon_2\zeta(k) + \|\mu_2 w(k)\|^2 + \mu_3^2 y_d \tag{4.26}$$

where $\hat{A}_d = [A_{d_1},\cdots,A_{d_N}]$, $\hat{B}_d = [B_{d_1},\cdots,B_{d_N}]$, $\hat{K} = \mathrm{diag}\{K,K,\cdots,K\}$, $X = A+BK$, $Y = \hat{A}_d + \hat{B}_d\hat{K}$ and

$$\Upsilon_2 = \begin{bmatrix} X^TPX + \sum_{i=1}^N S_i - P + \lambda^2 U^T U & X^TPY & X^TPF & X^TPE & X^TPH \\ Y^TPX & Y^TPY - \hat{S} & Y^TPF & Y^TPE & Y^TPH \\ F^TPX & F^TPY & F^TPF - \lambda^2 I & F^TPE & F^TPH \\ E^TPX & E^TPY & E^TPF & E^TPE - \mu_1^2 I & E^TPH \\ H^TPX & H^TPY & H^TPF & H^TPE & H^TPH - \mu_2^2 I \end{bmatrix}.$$

By pre-multiplying $\mathrm{diag}\{P,\cdots,P,I,I,I,P,I,I\}$ and post-multiplying $\mathrm{diag}\{P,\cdots,P, I,I,I,P,I,I\}$ on both sides of LMI (4.23), based on the Schur complement formula, it can be seen $\Upsilon_2 < \mathrm{diag}[-\mu_3^2 T,0,0,0,0]$ and for any $w(k)$ satisfying $\|w(k)\|_\infty \leq 1$, we can get

$$\Delta V(x(k),k) \leq -\mu_3^2 x^T(k)Tx(k) + \mu_1^2 + \mu_2^2 y_d. \tag{4.27}$$

Similarly to the proof of Theorem 4.1, it can be seen that inequality (4.19) still holds for the closed-loop system, which implies that system (4.22) is still stable in the presence of $w(t)$ and $V_g$. Meanwhile, by pre-multiplying $\mathrm{diag}\{T,I,I\}$, $\mathrm{diag}\{I,T\}$ and post-multiplying $\mathrm{diag}\{T,I,I\}$, $\mathrm{diag}\{I,T\}$ on both sides of LMIs (4.24) and (4.25), it is easy to show that LMIs (4.24) and (4.25) are equivalent to LMIs (4.14) and (4.15) so that the closed-loop system (4.22) still satisfies the peak-to-peak disturbance attenuation performance.

For a couple of $w(k)$ and $V_g$, we suppose that $\theta_1(k)$ and $\theta_2(k)$ are two trajectories of the closed-loop system (4.22) corresponding to a fixed initial condition. Defining $\sigma(k) := \theta_1(k) - \theta_2(k)$, then the dynamics for $\sigma(k)$ can be described by

$$\sigma(k+1) = (A+BK)\sigma(k) + \sum_{i=1}^N (A_{d_i} + B_{d_i}K)\sigma(k-d_i)$$

$$+F(f(\theta_1(k)) - f(\theta_2(k))). \tag{4.28}$$

Similarly to Equation 4.16, a Lyapunov function can be constructed as

$$V(\sigma(k),k) = \sigma^T(k)P\sigma(k) + \sum_{i=1}^N \sum_{l=1}^{d_i} \sigma^T(k-l)S_i\sigma(k-l)$$

$$+\sum_{i=1}^{k-1}[\|\lambda U\sigma(i)\|^2-\|\lambda(f(\theta_1(i))-f(\theta_2(i)))\|^2]. \tag{4.29}$$

Inequality (4.23) implies that

$$\begin{bmatrix} -P+\sum_{i=1}^{N}S_i+\mu_3^2T+\lambda^2U^TU & 0 & 0 & X^TP \\ 0 & -\hat{S} & 0 & Y^TP \\ 0 & 0 & -\lambda^2 I & F^TP \\ PX & PY & PF & -P \end{bmatrix} < 0.$$

Hence it can be seen that

$$\Delta V(\sigma(k),k) \leq -\mu_3^2\sigma^T(k)T\sigma(k) \leq -\mu_3^2\lambda_{\min}(T)\|\sigma(k)\|^2 < 0 \tag{4.30}$$

where $\lambda_{\min}(T)$ denotes the minimal eigenvalue of $T$. It can be verified that $\sigma = 0$ is the unique asymptotically stable equilibrium point of the system (4.28). This means that the closed-loop system (4.22) also has a unique asymptotically stable equilibrium point $x^* = [V^{*T}, \xi^{*T}]^T$. Thus, it can be verified that $\lim_{k\to\infty}W_e(k) = \lim_{k\to\infty}T_0^{-1}[\xi(k+1)-\xi(k)] = T_0^{-1}[\xi^*-\xi^*] = 0$, which shows that tracking performance can be achieved. □

### 4.3.3 Constrained Peak-to-peak Tracking Control

It has been shown that due to the property of the PDFs, the weight vectors considered have to satisfy $V^T(k)\Pi_0V(k) \leq 1$, which can be reduced to $x^T(k)\Pi x(k) \leq 1$, where $\Pi := \text{diag}\{\Pi_0, 0\} \geq 0$. Based on the property of the non-negative definite matrix, $\Pi$ can be divided into $\Pi = G^2$, where $G \geq 0$. Theorem 4.2 shows that the equilibrium of system (4.22) is not the origin. The existence of nonlinear dynamics and disturbances make the constrained tracking problem much more complicated, compared with the previous results (see [63, 64]).

**Theorem 4.3** For the known parameters $\lambda$, $\mu_i (i = 1,2,3)$, $\alpha > 0$ and matrix $U$, $G$, suppose that there exist $M = T^{-1} > 0$, $Q > 0$, $R$, $S_i > 0$ $i = 1,\cdots,N$ and parameter $\gamma > 0$ such that LMIs (4.23-4.25) and

$$\begin{bmatrix} M & MG \\ GM & \mu_3^2(\mu_1^2+\mu_2^2\gamma_d)^{-1}I \end{bmatrix} \geq 0, \tag{4.31}$$

and

$$\begin{bmatrix} M & MG \\ GM & I \end{bmatrix} \geq 0, \quad \begin{bmatrix} 1 & x_m^T \\ x_m & M \end{bmatrix} \geq 0 \tag{4.32}$$

are solvable, then closed-loop system (4.22) is stable, satisfies $x^T(k)\Pi x(k) \leq 1$, $\lim_{k\to\infty}V(k) = V_g$ and $\sup_{0\leq\|w\|\leq\infty}\frac{\|z(k)\|_\infty}{\|w(k)\|_\infty} < \gamma$. In this case, the PI control gain $K$ can be solved via $R = QK^T$.

*Proof.* Based on Theorem 4.2 it can be shown that the stability performance, tracking and disturbance attenuation of the closed-loop system can be guaranteed. It remains to show that under conditions (4.31) and (4.32), the closed-loop system satisfies the desired constraint with respective to PDF control.

Similarly to the above proof, inequality (4.19) still holds. This means that $x^T(k)Tx(k) \leq x_m^T T x_m$ or $x^T(k)Tx(k) \leq \mu_3^{-2}(\mu_1^2 + \mu_2^2 y_d)$. Combining with LMI (4.32), $\Pi \leq T$ and $x_m^T T x_m \leq 1$ can be satisfied, and we obtain

$$x^T(k)\Pi x(k) \leq x^T(k)Tx(k) \leq x_m^T T x_m \leq 1. \tag{4.33}$$

On the other hand, from LMI (4.31), we have $\Pi \leq \mu_3^2(\mu_1^2 + \mu_2^2 y_d)^{-1}T$, and it can be seen that

$$x^T(k)\Pi x(k) \leq \mu_3^2(\mu_1^2 + \mu_2^2 y_d)^{-1} x^T(k)Tx(k) \leq 1. \tag{4.34}$$

Based upon inequalities (4.33) and (4.34), it can be seen that constraint condition $x^T(k)\Pi x(k) \leq 1$ is satisfied. □

*Remark 4.3* Using the LMI-toolbox in MATLAB, the design procedures for the PI controller can be given as follows: (1) solve the LMIs (4.23-4.25), (4.31) and (4.32) for $M > 0$, $Q$, $R$, $S_i > 0$; (2) the PI control gain $K$ is solved via $R = QK^T$ and the control input can be provide by Equation 4.10. The above result shows that the design procedure can be reduced to a class of LMI algorithms with respect to $Q$ and $R$, which is more beneficial than the previous results in [61, 63, 64, 157]. For example, the algorithms provided in [61, 63, 64] were required to solve a class of complex matrix inequalities, which led to a non-convex computation procedure.

## 4.4 Simulations

Suppose that the output PDF can be approximated using the following B-spline models described by Equation 4.1 with $n = 3$, $y \in [0, 1.5]$, $i = 1, 2, 3$:

$$B_i(y) = \begin{cases} |\sin 2\pi y| & y \in [0.5(i-1); 0.5i] \\ 0 & y \in [0.5(j-1); 0.5j] \quad i \neq j. \end{cases}$$

Corresponding to Equation 4.3, it can be verified that

$$\Pi_0 = \int_0^{1.5} C_0^T(y)C_0(y)dy = \begin{bmatrix} \int_0^{1.5} B_1^2(y)dy & 0 \\ 0 & \int_0^{1.5} B_2^2(y)dy \end{bmatrix} = \begin{bmatrix} 0.25 & 0 \\ 0 & 0.25 \end{bmatrix},$$

and $h(V(k))$ is given by

$$h(V(k)) = \frac{\sqrt{-4[V^T(k)[\int_0^{1.5} C_0^T(y)C_0(y)dy]V(k) - 1]\int_0^{1.5} B_3^2(y)dy}}{2\int_0^{1.5} B_3^2(y)dy}$$

$$= \sqrt{4 - [v_1^2(k) + v_2^2(k)]}.$$

The desired PDF $g(y)$ is assumed to be described by Equation 4.8 with $V_g = [0.8, 1.1]^T$, and the dynamical relations between $V(k)$ and $u(k)$ is described by Equation 4.6 with the following selections:

$$A_0 = \begin{bmatrix} -1 & 0.5 \\ -0.5 & -1 \end{bmatrix}, A_{0d} = \begin{bmatrix} 0.5 & 0 \\ 0 & 0.5 \end{bmatrix}, F_0 = \begin{bmatrix} -0.3 & -0.3 \\ 0 & -0.3 \end{bmatrix}, B_0 = \begin{bmatrix} -0.5 & 0 \\ 0 & -0.5 \end{bmatrix},$$

$$B_{0d} = \begin{bmatrix} 0.3 & 0 \\ 0 & 0.3 \end{bmatrix}, E_0 = \begin{bmatrix} -0.5 & 0 \\ 0 & -0.5 \end{bmatrix}, T_0 = \begin{bmatrix} 0.2 & 0 \\ 0 & 0.2 \end{bmatrix}, f(V(k)) = \begin{bmatrix} v_2 - cosv_1 \\ sinv_2 \end{bmatrix}.$$

Defining $\lambda_1 = 2$, $\mu_1^2 = \mu_2^2 = \mu_3^2 = 2$, $\gamma = 6$ and solving the LMIs (4.23-4.25), (4.31) and (4.32), we obtain

$$K_P = \begin{bmatrix} -1.1971 & 0.8020 \\ -0.6719 & -1.0137 \end{bmatrix}, \quad K_I = \begin{bmatrix} 3.8274 & -0.4260 \\ 1.0854 & 4.7320 \end{bmatrix}.$$

Figure 4.1 displays the designed B-spline basis function. When the robust PI control law is applied, the responses of the dynamical weight vector are shown in Figure 4.2. Figure 4.3 is the trajectory of the PI control input. Under the proposed robust control strategy, the 3D mesh plot of the output PDF is showed in Figure 4.4. Figure 4.1-4.4 demonstrate that satisfactory tracking performance, stability and robustness have been achieved.

## 4.5 Conclusions

In this chapter we consider the PDF tracking control problem for general stochastic systems using square root B-spline expansions and a nonlinear weight dynamical model. A generalized PI control strategy is proposed in the discrete-time context. It is noted that the PDF tracking control problem can be reduced to a nonlinear weight tracking problem with state constraint. Convex LMI algorithms have been obtained such that the performances of stability, tracking and robustness as well as the state constraint are guaranteed simultaneously. Furthermore, this result also generalizes some previous work on classical constrained PI tracking control of nonlinear discrete-time systems.

Moreover, it should be pointed out that discrete-time system (4.6) can be generalized to sophisticated cases, for example, a system with uncertain parameters, time-varying delays and stochastic nonlinearities. This will be considered as a future research topic.

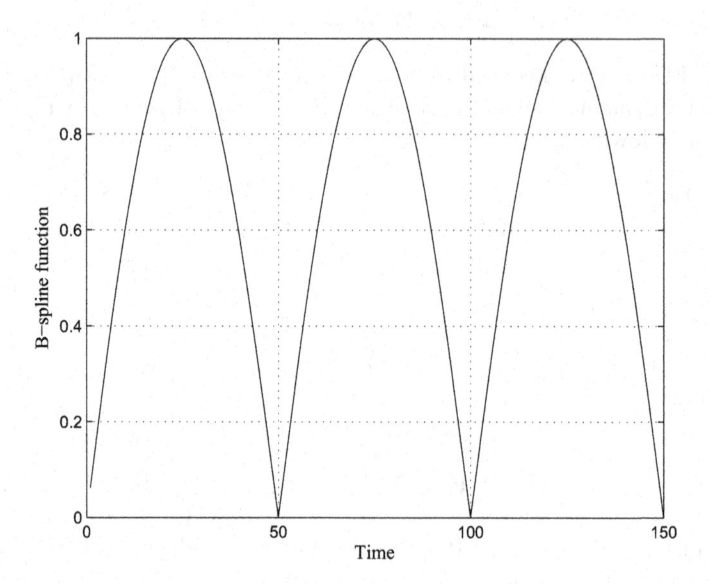


**Figure 4.1** Selected B-spline basis function

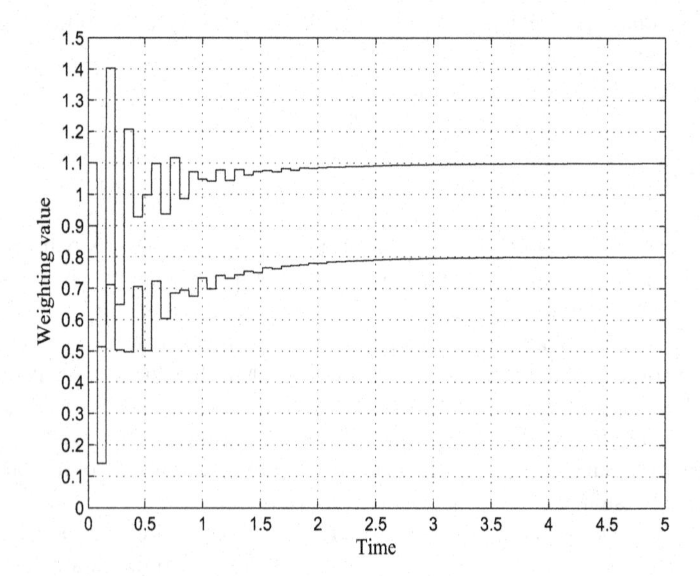


**Figure 4.2** Responses of the weight vector

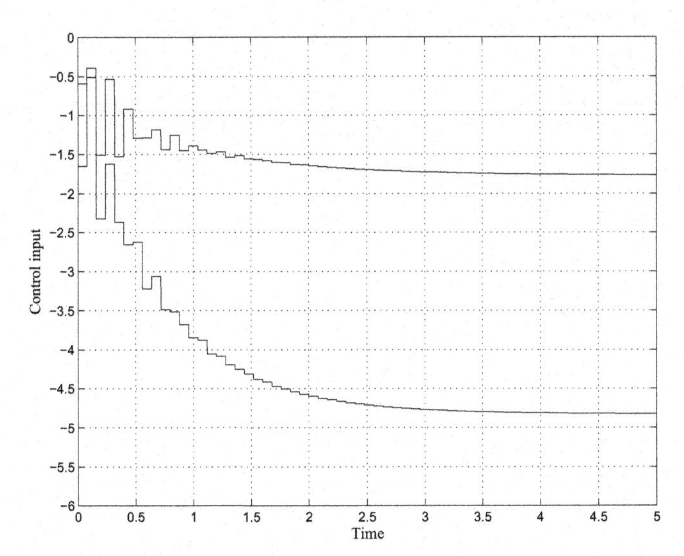


**Figure 4.3** PI type control input

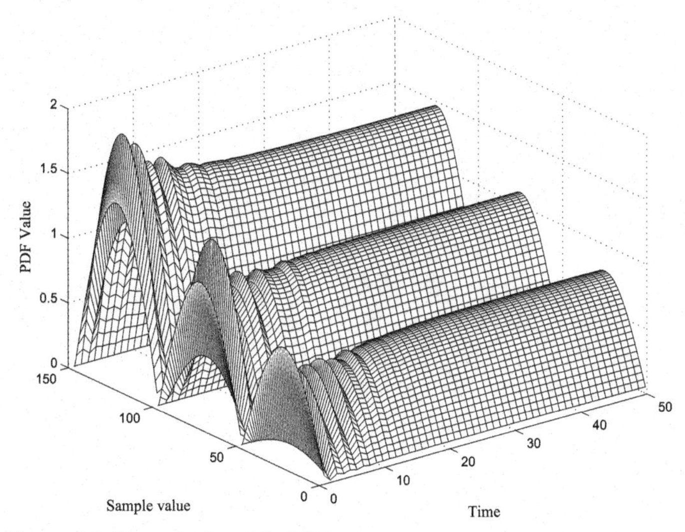


**Figure 4.4** 3D mesh plot of the PDF

# Part II
# Two-step Intelligent Optimization Modeling and Control for Stochastic Distribution Control Systems

Control and filtering for Gaussian systems have received much attention in past decades (see [3, 7, 18, 54, 180]), where mean and variance are two control objectives. However, non-Gaussian variables exist in many complex stochastic systems due to the nature of the random input sources and system nonlinearity. Indeed, these non-Gaussian variables may even have asymmetric and multiple-peak stochastic distributions (see [66, 157] for details). For non-Gaussian systems, it has been shown that mean and variance are insufficient to characterize the stochastic properties in terms of signal processing. As such, motivated by several typical examples in practical systems such as paper and board making (see [157, 161]), a group of new strategies that control the shape of output PDFs for general stochastic systems have been developed for both static and dynamic stochastic systems [24, 25, 37–39, 92, 134, 136, 141], [31, 58, 61–69, 72, 73, 87, 88, 109, 146, 155–169, 172, 173, 181–184, 186–188, 196]. These novel control approaches have been established in [63, 65] together with system matrix analysis and synthesis frameworks and have thus been called SDC. In order to obtain some feasible design algorithms and achieve the control objective, B-spline NN have been introduced to model the output PDFs so that the problem can be reduced to a tracking control problem for the weight systems. However, due to lack of model knowledge, most published results only address linear precise models (see [61, 63, 64, 157, 161, 164]) that obviously cannot satisfy industrial demands. Furthermore, some nonlinear models discussed (see [58, 66, 73, 109, 183, 184, 196]) were also difficult to obtain through traditional identification approaches.

Recently, due to massive parallelism and fast adaptability, NN based approximation and identification have been effectively used in many control problems for complex dynamic systems [49, 50, 75, 77, 118]. Generally, there are two well-known types of NN, namely static (feed-forward) NN and dynamic (recurrent) NN. Static NN (see [49–51, 74, 103, 194]) are usually used to approximate the nonlinear functions on the right-hand side of dynamic model equations because of their universal approximation property and learning capability. However, due to their slow learning rate and high sensitivity, most static NN have difficulties in acting as a black box identifier. Instead, dynamic neural networks (DNNs) (see [36, 96, 105, 123, 124, 126, 127, 130, 131, 189–191]) have some memory ability and can therefore capture system dynamics using measured data, which makes them much suited to identify complex nonlinear systems. [130] and [96] studied high order dynamic networks and multi-layer dynamic networks. In [126, 131, 189], on-line identification for nonlinear systems with adaptive laws via DNNs was studied, while [127, 130] investigated a control law based on DNNs so as to achieve the required tracking control processes. In [190], the passivity approach was applied to access several stability properties of the neuro identifier. A robust adaptive observer was designed to fulfill system identification in [16]. Also, [105] and [36] analyzed the control problem of an induction servomotor and a chaotic system, respectively via DNNs. [191] discussed the global asymptotic stability of recurrent NN with multiple time-varying delays.

Similarly, fuzzy techniques have also been widely and successfully used in nonlinear system modeling and control [170, 171]. Among various kinds of fuzzy methods, the well-known Takagi-Sugeno (T-S) fuzzy model [149] was recognized as a popular and powerful tool for approximating a complex nonlinear system. Recently, the analysis and synthesis of T-S fuzzy models have involved more complex nonlinear systems, for example, a descriptor system [148], time delay system [17, 174, 192], stochastic system [174] and networked control system [192]. Meanwhile, various techniques have also been developed for stability analysis of T-S fuzzy systems [94, 147, 174, 192]. On the other hand, the nonlinear tracking control problem has been considered through T-S fuzzy model as well [101, 107, 150, 177, 198]. In [177], a feedback linearization technique was proposed to design a fuzzy tracking controller. The variable structure control (VSC) approach has been applied to solve tracking problems for T-S fuzzy systems [198]. A simple observer-based fuzzy control technique was developed in [150] to reduce the tracking errors in the LMI approach.

In this part, both a NN and T-S fuzzy model, as system identifiers, are applied to solve the modeling and control problems in SDC. A new two-step NN model and fuzzy-neural model including both static and dynamic parts are established to accomplish the distribution control problem. It is noted that the new control framework not only represents a significant extension to the previous SDC results, but also has potential applications to general fuzzy/NN tracking control design problems.

# Chapter 5
# Adaptive Tracking Stochastic Distribution Control for Two-step Neural Network Models

## 5.1 Introduction

In this chapter, two-step NN are applied to solve the SDC problem for non-Gaussian systems (see Figure 5.1 for details). The objective is to control the conditional PDF of the system output to follow a given target function. B-spline NN are used to approximate the PDF of the system output, and DNNs are applied to identify the nonlinear relationships between the control input and the weight vectors. To achieve the control objective, a dynamic adaptive controller is developed so that the weight dynamics can follow the outputs of a reference model, as the desired weight values. Stability analysis for both the identification and tracking errors is developed via the use of the Lyapunov stability criterion. The main contributions of this chapter include:

1. Two kinds of NN models are first applied at the same time to solve the above mentioned stochastic control problem. Under the new control framework, it is much more intuitionist to design the control input and analyze the rigorous stability than previous results (see [61, 63, 64, 157]).

2. We attempt to import DNNs with undetermined parameters to perform black box identification in SDC. This represents a significant extension to the previous results [31, 58, 61–69, 72, 73, 87, 88, 109, 146, 155–169, 172, 173, 181–184, 186–188, 196].

3. Compared with other results related to DNNs [126, 131, 189], we consider the error term and construct the compensation term with a saturation function to guarantee the identification error convergence to zero.

4. Compared with references [63, 64, 157, 164], this chapter not only solves the dynamical tracking problem through the designed control input, but also identifies the dynamic trajectory of the weighting vectors related to the output PDFs.

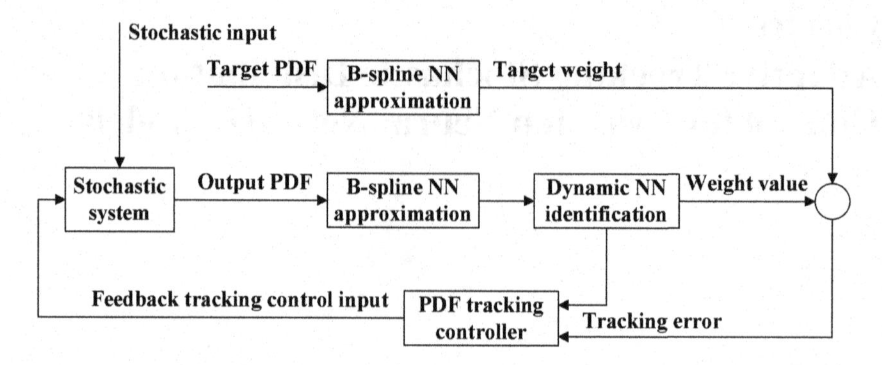


**Figure 5.1** PDF tracking control for SDC systems using two-step neural networks

## 5.2 Output PDF Model Using B-spline Neural Network

For a dynamic stochastic system, denote $u(t) \in R^m$ as the control input, $z(t) \in [a,b]$ as the stochastic output and the probability of output $z(t)$ lying inside $[a,y]$ can be described by

$$P(a \le z(t) < y, u(t)) = \int_a^y \gamma(\eta, u(t))\mathrm{d}\eta \tag{5.1}$$

where $\gamma(y,u(t))$ is the PDF of the stochastic variable $y(t)$ under control input $u(t)$. In this chapter, the following square root B-spline model will be adopted:

$$\sqrt{\gamma(y,u(t))} = \sum_{i=1}^{n} \upsilon_i(u(t))B_i(y) \tag{5.2}$$

where $B_i(y)(i = 1,\cdots,n)$ are pre-specified basis functions and $\upsilon_i(t) := \upsilon_i(u(t)),(i = 1,\cdots,n)$ are the corresponding weights. Due to $\int_a^b \gamma(y,u(t))\mathrm{d}y = 1$, only $n-1$ weights are independent. Equation 5.2 can be rewritten as

$$\sqrt{\gamma(y,u(t))} = C_0(y)V(t) + \upsilon_n(t)B_n(y) \tag{5.3}$$

where

$$C_0(y) = [B_1(y) \;\; B_2(y) \cdots \;\; B_{n-1}(y)],$$
$$V(t) = [\upsilon_1(t) \;\; \upsilon_2(t) \;\; \cdots \;\; \upsilon_{n-1}(t)]^T.$$

Denote

$$\Lambda_1 = \int_a^b C_0^T(y)C_0(t)\mathrm{d}y, \;\; \Lambda_2 = \int_a^b C_0(y)B_n(y)\mathrm{d}y, \;\; \Lambda_3 = \int_a^b B_n^2(y)\mathrm{d}y. \tag{5.4}$$

To guarantee $\int_a^b \gamma(y,u(t))\mathrm{d}y = 1$, we assume that $\{\upsilon_i(t) : i = 1,2,\cdots,n-1\}$ are independent, then it can be seen that $\upsilon_n(t)$ can be represented by $V(t)$, $C_0(y)$ and $B_n(y)$. Under conditions (5.3) and (5.4), $\upsilon_n$ can be represented as follows:

$$v_n(t) = h(V(t)) = \frac{\sqrt{\Lambda_3 - V^T(t)\Lambda_0 V(t)} - \Lambda_2 V(t)}{\Lambda_3}. \tag{5.5}$$

For $h(V(t))$, it is assumed that the Lipschitz condition can be satisfied within its operating region (see [66] for the details).

## 5.3 Dynamic Neural Network Identification

Once B-spline expansions have been made for the PDFs, the next step is to find the dynamic relationships between the input and the weights, corresponding to a further modeling procedure. It is noted that the nonlinear dynamics are difficult to obtain through traditional identification approaches. This is particularly true when a lot of data are available without complete model information. In the following, we will provide a DNN model to characterize the unknown weight dynamics, with a learning strategy for the model parameters. It is assumed that there exist optimal model parameters $W_1^*$, $W_2^*$ such that the nonlinear dynamics between the input $u(t)$ and the weights $V(t)$ can be described by the following NN model:

$$\begin{cases} \dot{x}(t) = Ax(t) + BW_1^*\sigma(x) + BW_2^*\phi(x)u(t) - F(t) \\ V(t) = Cx(t) \end{cases} \tag{5.6}$$

where $x \in R^m$ is the measurable state vector, $A$ is a stable matrix, $B \in R^{m\times m}$ is a diagonal matrix, of the form $B = \mathrm{diag}[b_1, b_2, \cdots, b_m]$. $C \in R^{(n-1)\times m}$ is a known matrix. $F(t)$ represents the error term and there exists an unknown positive constant $d$ such that $\|F(t)\|_1 \le d$. Obviously the optimal model parameters $W_1^*$, $W_2^*$ are bounded unknown matrices and satisfy $W_1^* W_1^{*T} \le \bar{W}_1$, $W_2^* W_2^{*T} \le \bar{W}_2$, where $\bar{W}_1$ and $\bar{W}_2$ are prior known matrices (see [189]).

The vector function $\sigma(x) \in R^m$ is assumed to be $m$-dimensional with elements increasing monotonically and the matrix function $\phi(x)$ is assumed to be $m \times m$ diagonal matrix. The typical presentation of the elements $\sigma_i(.), \phi_i(.)$ are as sigmoid functions, i.e.,

$$\sigma_i(x_i) = \frac{a}{1 + \mathrm{e}^{-bx_i}} - c. \tag{5.7}$$

The next step is to construct the following DNNs for identification

$$\begin{cases} \dot{\hat{x}}(t) = A\hat{x}(t) + BW_1\sigma(\hat{x}) + BW_2\phi(\hat{x})u(t) + u_f(t) \\ \hat{V}(t) = C\hat{x}(t) \end{cases} \tag{5.8}$$

where $\hat{x}(t) \in R^m$ is the state of the DNNs, $W_1 \in R^{m\times m}$ is the synaptic weights of the DNNs and $W_2$ is an $m \times m$ diagonal matrix of synaptic weights, of the form $W_2 = \mathrm{diag}[w_{21}, w_{22}, \cdots w_{2m}]$. $u_f(t)$ is the compensation term of the model error and will be defined later.

Denoting $\tilde{W}_1 = W_1 - W_1^*, \tilde{W}_2 = W_2 - W_2^*, \tilde{\sigma} = \sigma(\hat{x}) - \sigma(x), \tilde{\phi} = \phi(\hat{x}) - \phi(x)$ and identification error $e(t) = \hat{x}(t) - x(t)$. Because $\sigma(.)$ and $\phi(.)$ are chosen as sigmoid functions, they satisfy the following Lipschitz property:

$$\tilde{\sigma}^T \tilde{\sigma} \leq e^T(t) D_\sigma e(t), \quad (\tilde{\phi} u(t))^T (\tilde{\phi} u(t)) \leq \bar{u} e^T(t) D_\phi e(t) \tag{5.9}$$

where $D_\sigma, D_\phi$ are known positive definite matrices and $u(t)$ satisfy $u^T(t)u(t) \leq \bar{u}$, where $\bar{u}$ is a known constant.

From Equations 5.6 and 5.8, We obtain the error equation

$$\begin{aligned} \dot{e}(t) &= Ae(t) + B\tilde{W}_1 \sigma(\hat{x}) + B\tilde{W}_2 \phi(\hat{x}) u(t) + u_f(t) \\ &+ BW_1^* \tilde{\sigma} + BW_2^* \tilde{\phi} u(t) + F(t). \end{aligned} \tag{5.10}$$

If we define $Q = D_\sigma + \bar{u} D_\phi + Q_0$, $R = B(\bar{W}_1 + \bar{W}_2)B^T$, there exist a stable matrix $A$ and a strictly positive definite matrix $Q_0$ such that the matrix Riccati equation

$$A^T P + PA + PRP = -Q \tag{5.11}$$

has a positive definite solution $P$.

The following result gives a stable learning procedure for DNN identification.

**Theorem 5.1** The compensation term is defined as $u_f(t) = -SGN\{Pe(t)\}\hat{d}$, and the parameters $W_1$, $W_2$, $\hat{d}$ are updated as

$$\begin{aligned} \dot{W}_1 &= -\gamma_1 BPe(t) \sigma^T(\hat{x}(t)), \\ \dot{W}_2 &= -\gamma_2 \Theta[BPe(t) u^T(t) \phi(\hat{x}(t))], \\ \dot{\hat{d}} &= \gamma_3 \|Pe(t)\|_1 \end{aligned} \tag{5.12}$$

where $\hat{d}$ are estimation values of unknown constants $d$, $\gamma_i (i = 1,2,3)$ are defined positive constants, $\Theta[.]$ represent a kind of transformation that make the common matrix into a diagonal matrix, and $P$ is the solution of the Riccati equation (5.11), then the error dynamics of the identification scheme described by Equation 5.10 satisfy $\lim_{t \to \infty} \|e(t)\| = 0$.

*Proof.* Select a Lyapunov function as

$$V(e(t),t) = e^T(t) P e(t) + \mathrm{tr}\{\tilde{W}_1^T \gamma_1^{-1} \tilde{W}_1\} + \mathrm{tr}\{\tilde{W}_2^T \gamma_2^{-1} \tilde{W}_2\} + \tilde{d}^T \gamma_3^{-1} \tilde{d} \tag{5.13}$$

where $P \in R^{m \times m}$ is a positive definite solution of Equation 5.11. According to Equation 5.10, the derivative of $V(e(t),t)$ can be formulated as

$$\begin{aligned} \dot{V}(e(t),t) &= e^T(t)(PA + A^T P)e(t) + 2e^T(t) PB\tilde{W}_1 \sigma(\hat{x}) + 2e^T(t) PB\tilde{W}_2 \phi(\hat{x}) u(t) \\ &+ 2e^T(t) P u_f(t) + 2e^T(t) PF(t) + 2e^T(t) PBW_1^* \tilde{\sigma} + 2e^T(t) PBW_2^* \tilde{\phi} u(t) \\ &+ 2\mathrm{tr}\{\dot{\tilde{W}}_1^T \gamma_1^{-1} \tilde{W}_1\} + 2\mathrm{tr}\{\dot{\tilde{W}}_2^T \gamma_2^{-1} \tilde{W}_2\} + 2\dot{\tilde{d}}^T \gamma_3^{-1} \tilde{d}. \end{aligned} \tag{5.14}$$

Since $e^T(t) PBW_1^* \tilde{\sigma}$ and $e^T(t) PBW_2^* \tilde{\phi} u(t)$ are scalars, it can be obtained that

$$2e^T(t)PBW_1^*\tilde{\sigma} \le e^T(t)PBW_1^*W_1^{*T}B^TPe(t) + \tilde{\sigma}^T\tilde{\sigma},$$

$$2e^T(t)PBW_2^*\tilde{\phi}u(t) \le e^T(t)PBW_2^*W_2^{*T}B^TPe(t) + (\tilde{\phi}u(t))^T\tilde{\phi}u(t). \tag{5.15}$$

Based on inequalities (5.9) and (5.15) , we have

$$\begin{aligned}\dot{V}(e(t),t) \le\ & e^T(t)[PA + A^TP + PB(\bar{W}_1 + \bar{W}_2)B^TP + (D_\sigma + \bar{u}D_\phi + Q_0)]e(t)\\ & +2\mathrm{tr}\{\dot{\tilde{W}}_1^T\gamma_1^{-1}\tilde{W}_1\} + 2e^T(t)PB\tilde{W}_1\sigma(\hat{x}) + 2\mathrm{tr}\{\dot{\tilde{W}}_2^T\gamma_2^{-1}\tilde{W}_2\}\\ & +2e^T(t)PB\tilde{W}_2\phi(\hat{x})u(t) - 2e^T(t)P\cdot SGN\{Pe(t)\}\hat{d} + 2d\|Pe(t)\|_1\\ & +2\dot{\tilde{d}}^T\gamma_3^{-1}\tilde{d} - e^T(t)Q_0e(t).\end{aligned} \tag{5.16}$$

Since $\dot{\tilde{W}}_1 = \dot{W}_1, \dot{\tilde{W}}_2 = \dot{W}_2, \dot{\tilde{d}} = \dot{\hat{d}}$ and using Riccati equation (5.11), it can be verified that

$$\begin{aligned}\dot{V}(e(t),t) \le\ & -e^T(t)Q_0e(t) + 2\mathrm{tr}\{\dot{W}_1^T\gamma_1^{-1}\tilde{W}_1\} + 2e^T(t)PB\tilde{W}_1\sigma(\hat{x})\\ & +2\mathrm{tr}\{\dot{W}_2^T\gamma_2^{-1}\tilde{W}_2\} + 2e^T(t)PB\tilde{W}_2\phi(\hat{x})u(t) - 2\|Pe(t)\|_1\tilde{d} + 2\dot{\tilde{d}}^T\gamma_3^{-1}\tilde{d}.\end{aligned}$$

So from the updating laws (5.12), we obtain

$$\dot{V}(e(t),t) \le -e^T(t)Q_0e(t) < 0. \tag{5.17}$$

Hence, $V \in L_\infty$, which implies $e(t) \in L_\infty$. From Equation 5.10, we have $\dot{e}(t) \in L_\infty$. Integrating both sides of inequality (5.17) yields

$$\int_0^\infty \|e(t)\|^2 \mathrm{d}t \le V(0) - V(\infty) < \infty. \tag{5.18}$$

So $e(t) \in L_2 \bigcap L_\infty$, using the Barbalat lemma [170, 171], we have $\lim_{t\to\infty}\|e(t)\| = 0$. □

*Remark 5.1* It is noted that the Riccati equation (5.11) can be replaced by the following linear matrix equation:

$$\begin{bmatrix} A^TP + PA + D_\sigma + \bar{u}D_\phi + Q_0 & PB \\ B^TP & -(\bar{W}_1 + \bar{W}_2)^{-1} \end{bmatrix} < 0 \tag{5.19}$$

where $P > 0$ is the solution of LMI (5.19). From a theoretical point of view, based on LMI (5.19), the identification error $e(t)$ is also convergent to zero.

## 5.4 Adaptive Tracking Control for Weight Reference Model

In this section, we investigate the tracking problem. Corresponding to Equation 5.3, a desired (known) PDF to be tracked can be described by

$$\sqrt{g(y)} = C_0(y)V_g + h(V_g)B_n(y) \tag{5.20}$$

where $V_g$ is the desired weight vector corresponding to $B_i(y)$. The tracking objective is to find $u(t)$ such that $\gamma(y,u(t))$ can follow $g(y)$. The error between the output PDF and the target PDF is formulated by $\Delta_e = \sqrt{g(y)} - \sqrt{\gamma(y,u(t))}$, i.e.,

$$\Delta_e = C_0V_e + [h(V_g) - h(V(t))]B_n(y) \tag{5.21}$$

where $V_e = V_g - V(t)$. Due to continuity of $h(V(t))$, $\Delta_e \longrightarrow 0$ holds as long as $V_e \longrightarrow 0$. At this stage, the PDF control problem can be formulated into the tracking problem for the above nonlinear weight systems.

In this stage, a dynamic reference model is considered as part of the target model. The desired weight vector $V_g \in R^{n-1}$ can be obtained by the following dynamic reference model:

$$\begin{cases} \dot{x}_m = A_m x_m + B_m r \\ V_m(t) = C_m x_m(t) \end{cases} \tag{5.22}$$

where $x_m \in R^m$ are the reference model states, $r \in R^m$ are inputs and $A_m, B_m, C_m$ are constant matrices of appropriate dimensions. In order to guarantee $\lim_{t\to\infty} V_m(t) = V_g$, parameter matrices $(A_m, B_m, C_m)$ should satisfy $V_g = C_m(I - A_m)^{-1}B_m r$ (see [63]).

In the following, the problem is transformed into a nonlinear dynamic control problem for tracking error vector $e_v(t) = V(t) - V_m(t)$. From Equations 5.8 and 5.22, the error $e_v(t)$ can be expressed as

$$e_v(t) = C\hat{x} - C_m x_m. \tag{5.23}$$

Define $\bar{C} = \begin{bmatrix} C \\ C_1 \end{bmatrix}$, $\bar{C}_m = \begin{bmatrix} C_m \\ C_{m1} \end{bmatrix}$, where $C_1 \in R^{(m-n+1)\times m}, C_{m1} \in R^{(m-n+1)\times m}$ are artribary matrices which satisfy $|\bar{C}| \neq 0$ and $|\bar{C}_m| \neq 0$. So we obtain

$$\bar{e}_v(t) = \bar{C}\hat{x} - \bar{C}_m x_m. \tag{5.24}$$

Furthermore, we have

$$\begin{aligned} \dot{\bar{e}}_v(t) &= \bar{C}A\hat{x} + \bar{C}BW_1\sigma(\hat{x}) + \bar{C}BW_2\phi(\hat{x})u(t) \\ &+ \bar{C}u_f(t) - \bar{C}_m A_m x_m - \bar{C}_m B_m r. \end{aligned} \tag{5.25}$$

Taking $u(t)$ to be given by

$$u(t) = -[\bar{C}BW_2\phi(\hat{x})]^{-1}[\bar{C}A\bar{C}^{-1}\bar{C}_m x_m + \bar{C}BW_1\sigma(\hat{x})$$

$$+\bar{C}u_f(t) - \bar{C}_m A_m x_m - \bar{C}_m B_m r] \tag{5.26}$$

and substituting this into Equation 5.25 we obtain

$$\dot{\bar{e}}_v(t) = \bar{C}A\bar{C}^{-1}\bar{e}_v(t) = \bar{A}\bar{e}_v(t). \tag{5.27}$$

For the matrix $\bar{A} = \bar{C}A\bar{C}^{-1}$, we assume that there exists a strictly positive definite matrix $Q_1$ such that the matrix Lyapunov equation

$$\bar{A}^T\bar{P} + \bar{P}\bar{A} = -Q_1 \tag{5.28}$$

has a positive definite solution $\bar{P}$.

In order to assure the existence of $[\bar{C}BW_2\phi(\hat{x})]^{-1}$, we need to establish $w_{2i} \neq 0$. Hence $W_2$ is confined through the use of modified adaptive projection algorithms. In particular, the standard adaptive laws are modified to

$$\dot{W}_1 = \begin{cases} -\gamma_1 BPe(t)\sigma^T(\hat{x}) \;\; when \; \|W_1\| < M_1 \;\; or \; \|W_1\| = M_1 \\ \qquad\qquad and \;\; \mathrm{tr}\{\sigma(\hat{x})e^T(t)PBW_1\} \geq 0 \\ -\gamma_1 BPe(t)\sigma^T(\hat{x}) + \gamma_1 \mathrm{tr}\{\sigma(\hat{x})e^T(t)PBW_1\}\frac{W_1}{\|W_1\|^2} \;\; when \;\; \|W_1\| = M_1 \\ \qquad\qquad and \;\; \mathrm{tr}\{\sigma(\hat{x})e^T(t)PBW_1\} < 0 \end{cases}. \tag{5.29}$$

When $w_{2i} = \varepsilon$, we adopt

$$\dot{w}_{2i} = \begin{cases} -\gamma_2 b_i u_i \phi_i(\hat{x})e^T(t)P_i \;\; when b_i u_i \phi_i(\hat{x})e^T(t)P_i < 0 \\ 0 \qquad when \;\; b_i u_i \phi_i(\hat{x})e^T(t)P_i \geq 0 \end{cases} \tag{5.30}$$

where $u_i$ is the $i$th element of $u(t)$ and $P_i$ is the $i$th column of $P$. Otherwise

$$\dot{W}_2 = \begin{cases} -\gamma_2\Theta[BPe(t)u^T(t)\phi(\hat{x}(t)))] \;\; when \;\; \|W_2\| < M_2 \;\; or \;\; \|W_2\| = M_2 \\ \qquad and \;\; \mathrm{tr}\{BPe(t)u^T(t)\phi(\hat{x}(t))W_2\} \geq 0 \\ -\gamma_2\Theta[BPe(t)u^T(t)\phi(\hat{x}(t))] + \gamma_2 \mathrm{tr}\{BPe(t)u^T(t)\phi(\hat{x}(t))W_2\}\frac{W_2}{\|W_2\|^2} \\ \qquad when \;\; \|W_2\| = M_2 \;\; and \;\; \mathrm{tr}\{BPe(t)u^T(t)\phi(\hat{x}(t))W_2\} < 0 \end{cases}, \tag{5.31}$$

$$\dot{\hat{d}} = \gamma_3\|Pe(t)\|_1. \tag{5.32}$$

□

**Theorem 5.2** Consider the DNNs (5.8), the dynamic reference model (5.22), the control law (5.26) and the adaptive law (5.29-5.32); we have the following properties:

1. $\|W_1\| \leq M_1; \|W_2\| \leq M_2, \;\; w_{2i} \geq \varepsilon$, where $M_1, M_2, \varepsilon$ are known positive constants and satisfy $M_1^2 > \mathrm{tr}(\bar{W}_1)$, $M_2^2 > \mathrm{tr}(\bar{W}_2)$.
2. $\lim_{t\to\infty} e(t) = \lim_{t\to\infty} e_v(t) = 0$, $\lim_{t\to\infty} V(t) = \lim_{t\to\infty} V_m(t) = V_g$.

*Proof.*

1. See [170, 171] for details.
2. Select a Lyapunov function as

$$V_1(e(t),t) = e^T(t)Pe(t) + \bar{e}_v^T \bar{P}\bar{e}_v + \text{tr}\{\tilde{W}_1^T \gamma_1^{-1} \tilde{W}_1\}$$
$$+\text{tr}\{\tilde{W}_2^T \gamma_2^{-1} \tilde{W}_2\} + \tilde{d}^T \gamma_3^{-1} \tilde{d} \tag{5.33}$$

where $P$ and $\bar{P}$ satisfy Equations 5.11 and 5.28, respectively.

From Equations 5.25-5.32, $\dot{V}_1(e(t),t)$ can be expressed as

$$\dot{V}_1(e(t),t) \le -e^T(t)Q_0 e(t) - \bar{e}_v^T Q_1 \bar{e}_v + I_1 \text{tr}\{\sigma(\hat{x})e(t)^T PBW_1\}\text{tr}\{\frac{W_1^T \tilde{W}_1}{||W_1||^2}\}$$

$$+I_2 \text{tr}\{BPe(t)u^T(t)\phi(\hat{x}(t))W_2\}\text{tr}\{\frac{W_2^T \tilde{W}_2}{||W_2||^2}\}$$

where $I_1 = 0(1)$, if the first (second) condition of Equation 5.29 is true. $I_2 = 0(1)$, if the first (second) condition of Equation 5.31 is true. When the second conditions of Equations 5.29 and 5.31 are true, this implies $||W_i^*|| \le \sqrt{\text{tr}(\bar{W}_i)} \le M_i = ||W_i||$. So for $i = 1,2$, we have

$$\text{tr}\{W_i^T \tilde{W}_i\} = \text{tr}\{W_i^T(W_i - W_1^*)\} = \frac{1}{2}\text{tr}[W_i^T W_i - W_i^{*^T} W_i^* + \tilde{W}_i^T \tilde{W}_i] \ge 0. \tag{5.34}$$

Furthermore we obtain

$$\dot{V}_1(e(t),t) \le -e^T(t)Q_0 e(t) - \bar{e}_v^T(t) Q_1 \bar{e}_v(t) \le 0. \tag{5.35}$$

Integrating both sides of inequality (5.35) yields

$$\int_0^\infty (||e(t)||^2 + ||\bar{e}_v(t)||^2)\text{d}t \le V_1(0) - V_1(\infty) < \infty. \tag{5.36}$$

Similarly to Theorem 5.1, using the Barbalat lemma, we obtain

$$\lim_{t\to\infty} e(t) = \lim_{t\to\infty} e_v(t) = \lim_{t\to\infty} \bar{e}_v(t) = 0, \quad \lim_{t\to\infty} V(t) = \lim_{t\to\infty} V_m(t) = V_g.$$

□

*Remark 5.2* It is noted that the modified adaptive laws (5.29-5.32) can satisfy our demands rather than those of [170, 171]. Equations 5.29 and 5.31 can guarantee the bound for weights $W_1$ and $W_2$. Define $\Phi_1$, $\Phi_2$ to be constraint sets for $W_1$ and $W_2$, respectively, that is $\Phi_i = \{W_i : \ \text{tr}(W_i W_i^T) \le M_i^2, M_i > 0\}, i = 1,2$. In Equation 5.31, $\Theta[.]$ is applied because $W_2$ is a diagonal matrix. The initial value of $w_{2i}$ is chosen to be larger than $\varepsilon$ so that $w_{2i} \ge \varepsilon$ can be ensured through Equation 5.30.

## 5.5 Illustrative Examples

### 5.5.1 Example 1

Suppose that the output PDFs can be approximated using the square root B-spline models described by Equation 5.3 with $n = 3$, $y \in [0, 1.5]$, $i = 1, 2, 3$

$$B_i(y) = \begin{cases} |\sin 2\pi y| & y \in [0.5(i-1); 0.5i] \\ 0 & y \in [0.5(j-1); 0.5j] \quad i \neq j. \end{cases}$$

From the notation in Equation 5.4, it can be seen that $\Lambda_1 = \mathrm{diag}\{0.25, 0.25\}, \Lambda_2 = [0, 0], \Lambda_3 = 0.25$. The desired PDF $g(y)$ is assumed to be described by $V_g = [\frac{\pi}{3}, \frac{\pi}{6}]^T$.

In the chapter, the nonlinear dynamic relationships between the input and the weights related to the PDF are assumed to be given by the difference equation

$$\dot{x}(t) = A_1 x(t) + B_1 f(x) + C_1 u(t) + d_1, \tag{5.37}$$

$$A_1 = \begin{bmatrix} -3 & 0 & -1 \\ 2 & -4 & -1 \\ 2 & 0 & -3 \end{bmatrix}, \quad C_1 = \begin{bmatrix} 1 & 0 & 0 \\ 0 & 1 & 0 \\ 0 & 0 & 1 \end{bmatrix},$$

$$x_0 = \begin{bmatrix} 2 \\ 3 \\ 0 \end{bmatrix}, B_1 = \begin{bmatrix} 1 \\ -1 \\ -1 \end{bmatrix}, d_1 = \begin{bmatrix} 0.5 \\ 0.5 \\ -0.5 \end{bmatrix},$$

$$f(x) = 2\sin x_1 - 6\cos x_2 + 2\sin x_3.$$

For the selected DNNs, we assume that $\sigma(x_i) = \phi(x_i) = \frac{2}{1+e^{-0.5x_i}} + 0.5$ and

$$\hat{x}_0 = \begin{bmatrix} 2 \\ 3 \\ 0 \end{bmatrix}, \quad A = \begin{bmatrix} -3 & 0 & -2 \\ 0 & -2 & -2 \\ 2 & 0 & -2 \end{bmatrix}, \quad C = \begin{bmatrix} \frac{2}{3} & 0 & 0 \\ 0 & \frac{1}{3} & 0 \end{bmatrix}.$$

In reference model (5.22),

$$A_m = \mathrm{diag}\{-2, -2, -2\}, \quad B_m = [1, -0.5, 1]^T,$$

$$X_{m,0} = [-2, -2, -1]^T, \quad C_m = [\frac{2\pi}{3}, 0, 0; 0, -\frac{2\pi}{3}, 0].$$

In adaptive laws (5.29-5.32), $\gamma_i = 3, i = 1, 2, 3, 4, \hat{d}(0) = 2$, $W_{1,0} = W_{2,0} = I$. It is noted that the responses of nonlinear system (5.37) are shown in Figure 5.2 and the states of DNN (5.8) are shown in Figure 5.3. Figure 5.4 and Figure 5.5 are the output trajectories of the DNN and reference model, respectively. The control law is shown in Figure 5.6. Figure 5.7 shows the 3D mesh plot of the output PDFs. Figsure 5.2-5.7 demonstrate that satisfactory tracking performance, identification capability and robustness have been achieved.

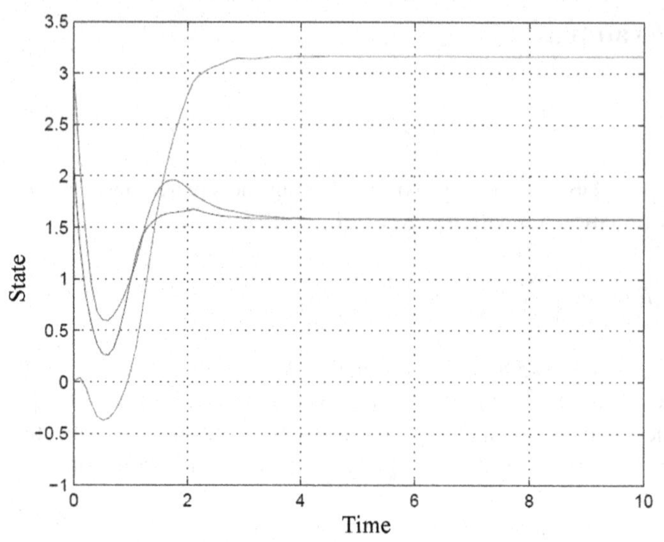


**Figure 5.2** Responses of unknown dynamics

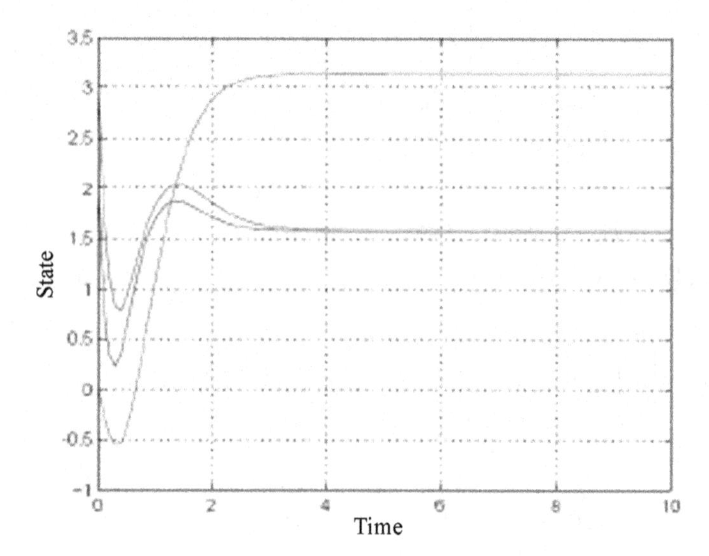


**Figure 5.3** Responses of the DNNs

## *5.5.2 Example 2*

A typical example of a paper machine is shown in Figure 5.8. In the head box, fiber, fillers and other chemical additives are mixed. This mixture generally consists of solids and water. When this mixture is injected onto the moving wire table, some water is drained through the wire nets into a white water pit underneath the wire

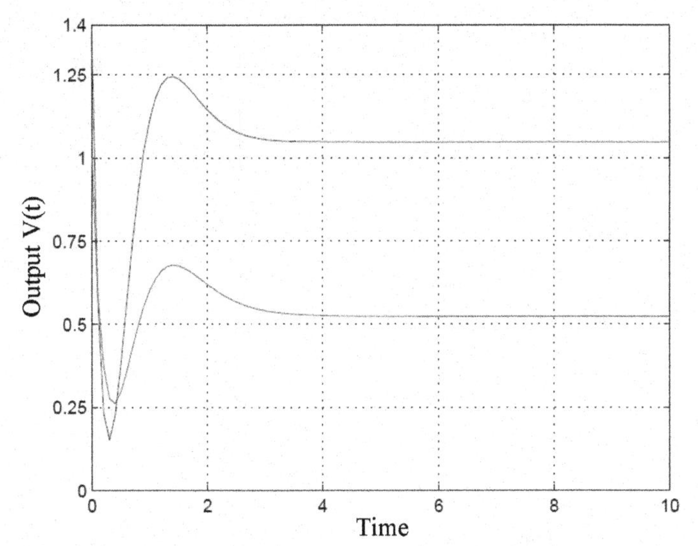


**Figure 5.4** Outputs of the DNNs

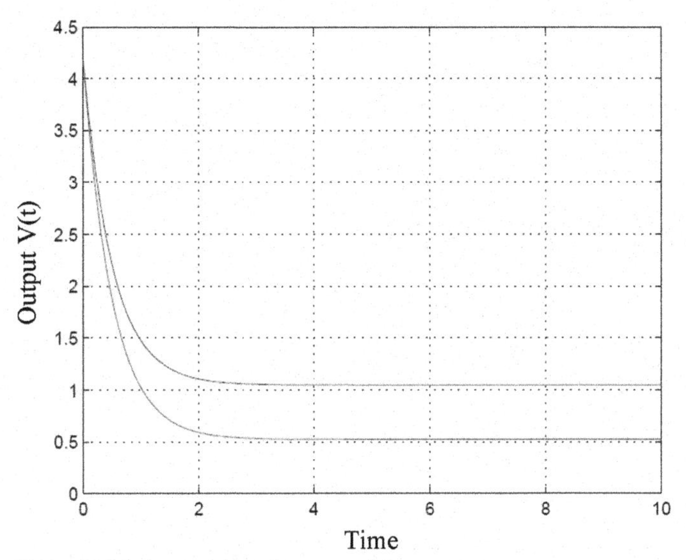


**Figure 5.5** Outputs of the reference model

table. This drainage process is continuous and the water in the pit also contains some solids in the form of either flocculation, or particles with a size distribution. As such, in order to control the efficiency of raw material usage, the total solids in the drained water need to be controlled and minimized.

As discussed in [139, 161], a group of chemicals, known as retention aids, can help to control such a solid distribution. As a result, flocculation will occur and the density of the solid distribution can be locally increased via increased retention

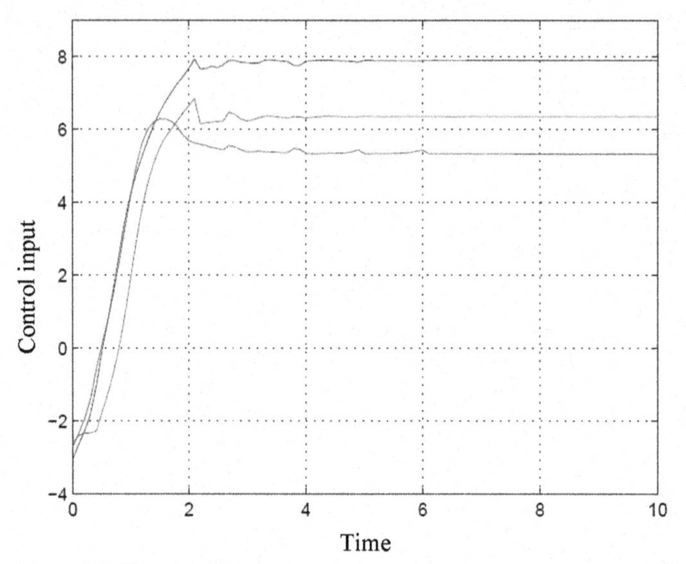


**Figure 5.6** The control input

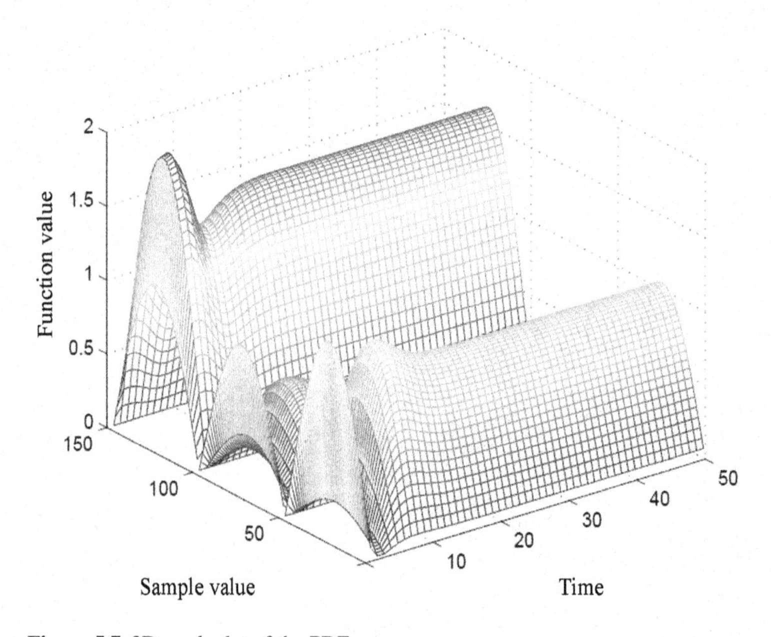


**Figure 5.7** 3D mesh plot of the PDF

aids. It can thus be concluded that these retention polymers are used to control (minimize) the size distribution of solid flocculation in the drained water. Indeed, since the head box can be regarded as a tank level system, at least a first order dynamic exists between the input (i.e., the retention aids) and the output distribution (i.e.,

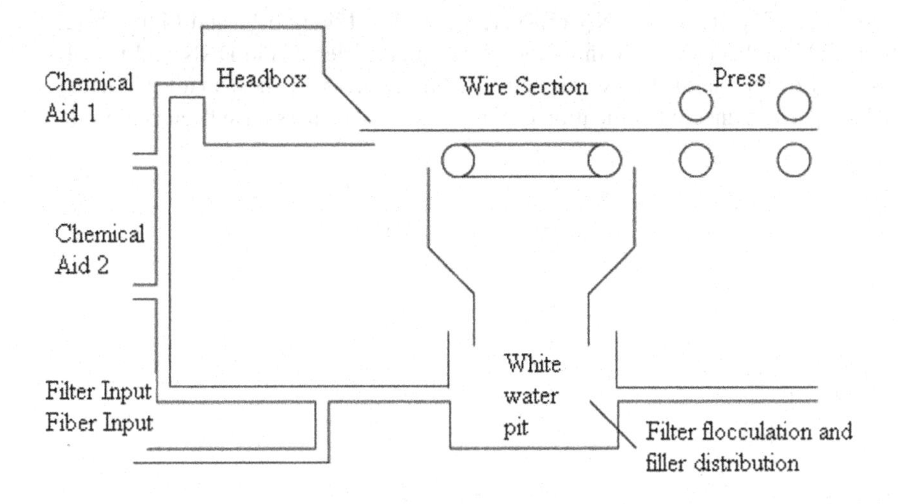


**Figure 5.8** A paper machine wet end

the flocculation size distribution in the white water pit). With reference to several laboratory tests on the flocculation size distribution in the white water system, 100 samples have been formulated and used to evaluate 100 different PDFs. The distribution of flocculation sizes can be approximated by a $\Gamma$-distribution under certain assumptions (see [139, 161] for details).

Since this distribution involves an exponential function, it is possible to use a set of exponential type spline functions to approximate the actual flocculation size distribution, given by the following:

$$B_i(y) = \exp(-(y-y_i)^2\sigma_i^{-2}) \quad (i = 1,2,\cdots,5)$$

where $V(t) = [v_1, v_2, v_3, v_4, v_5]^T$, $y \in [0, 0.05], y_i = 0.003 + 0.006(i-1), \sigma_i = 0.03$ $(i = 1,2,\cdots,5)$. For simplicity, the desired PDF $g(y)$ is assumed to be described by $V_g = [0.5, 0.6, 0.7, 0.8]^T$. The dynamic model between $u(t)$ and $x(t)$ is described by DNNs (5.8). Similar to the modeling procedures of [9], the coefficient matrices are denoted by $\sigma(x_i) = \phi(x_i) = \frac{2}{1+e^{-0.5x_i}} + 0.5 \;\; i = 1,2,3,4,5$.

$$A = \text{diag}\{-0.083, -0.083, -0.083, -0.083, -0.083\},$$

$$B = \text{diag}\{0.78, 0.78, 0.78, 0.78, 0.78\},$$

$$C = [-1,0,1,0,1;0,1,-1,0,0;1,0,0,0,1;0,-1,1,1,1].$$

For the unknown dynamic model (5.38), we choose $A_1 = \text{diag}\{1;-1;1;-1;1\}$, $B_1 = I_5$, $C_1 = d_1 = \text{diag}\{0.2;0.2;0.2;0.2;0.2\}$, and $f(x) = [\sin(3x_1 - 2x_2), \sin 2x_2, \cos(2x_3 - x_2), 2\cos x_4, 3\sin x_5]^T$. The adaptive laws and other parameters are the same as Example 1. The responses of nonlinear system (5.38) are shown in Figure 5.9 and

Figure 5.10. The states of DNN (5.8) are showed in Figure 5.11 and Figure 5.12. Figure 5.13 and Figure 5.14 show the output trajectories of the DNN and the 3D mesh plot of the output PDFs, respectively. Figure 5.9-5.14 show that satisfactory tracking performance, identification capability and robustness have been achieved.

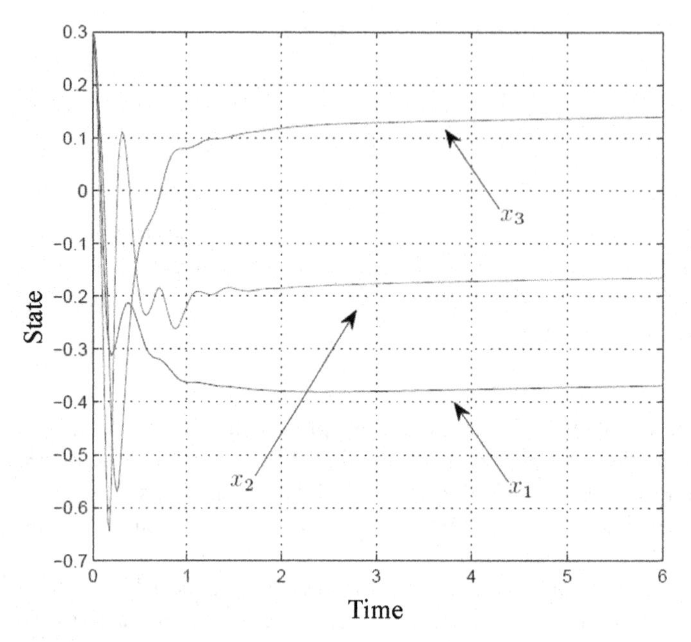


**Figure 5.9** Responses of unknown dynamics

## 5.6 Conclusions

The chapter describes the use of two-step NN to solve the tracking control problem for general non-Gaussian stochastic systems. After B-spline approximation to the measured output PDFs, DNNs can be used to describe the complex nonlinear relationships between control input and weight vectors related to output PDF. Using adaptive projection algorithms, an adaptive state feedback controller can be obtained such that the closed-loop tracking performance, system identification and robustness are guaranteed simultaneously. Simulations are described to demonstrate the efficiency of the proposed approach.

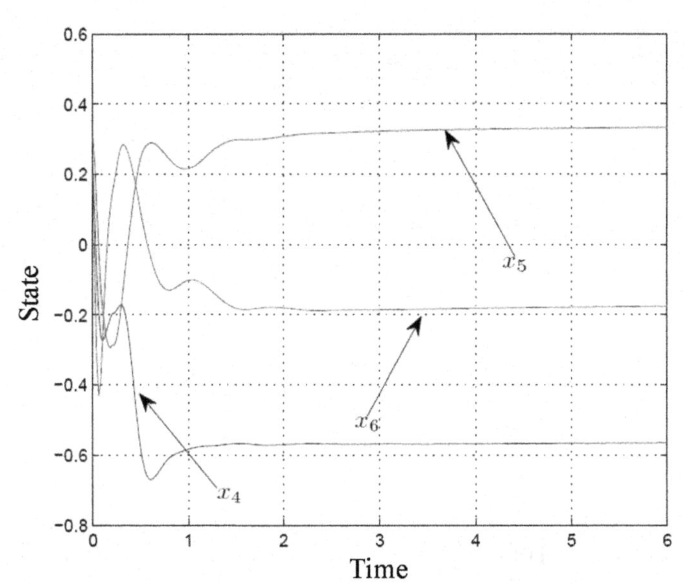


**Figure 5.10** Responses of unknown dynamics

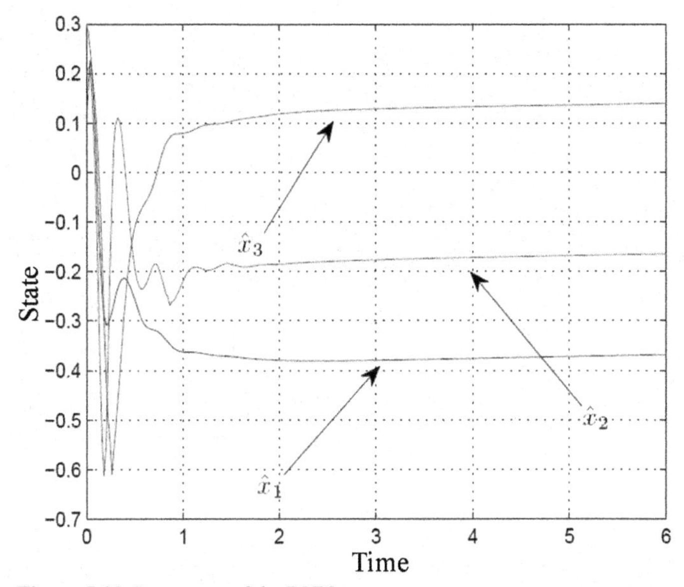


**Figure 5.11** Responses of the DNNs

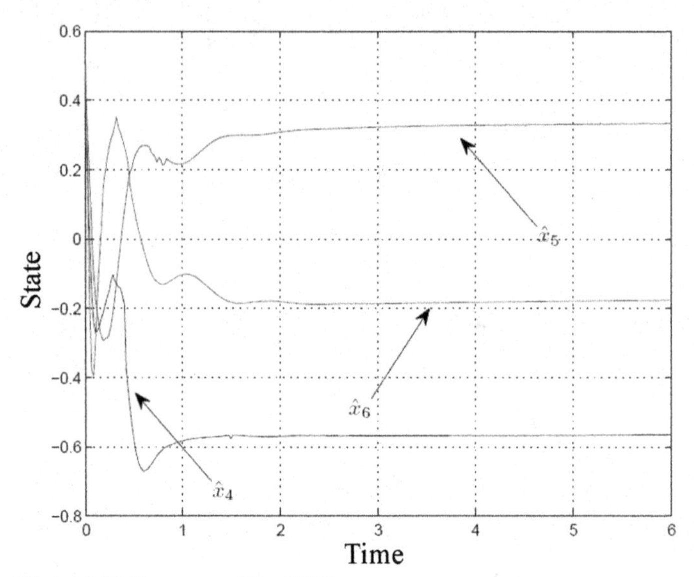


**Figure 5.12** Responses of the DNNs

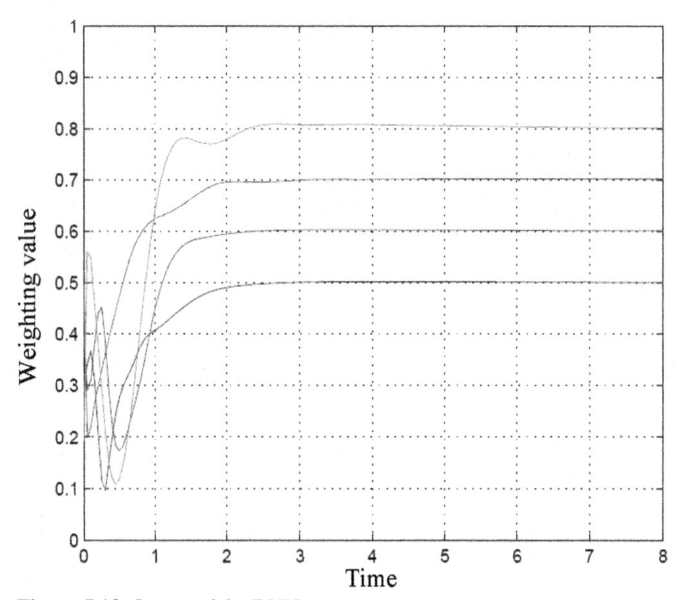


**Figure 5.13** Output of the DNNs

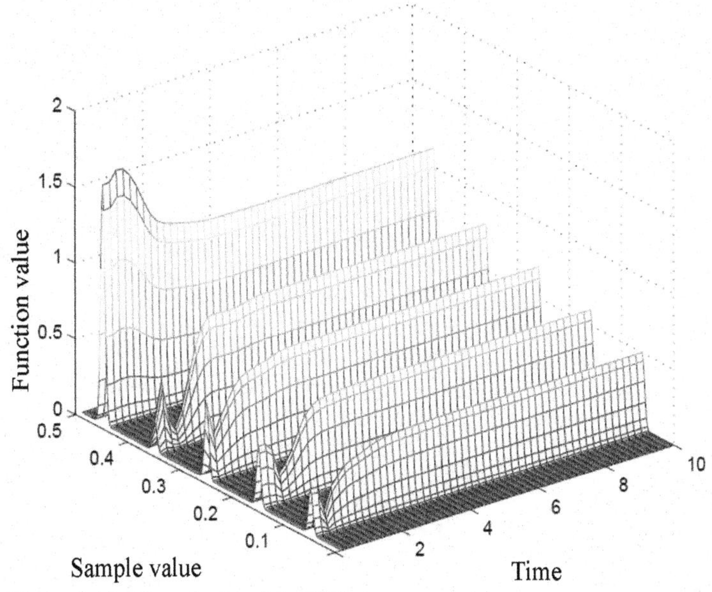


**Figure 5.14** 3D mesh plot of the PDF

# Chapter 6
# Constrained Adaptive Proportional Integral Tracking Control for Two-step Neural Network Models with Delays

## 6.1 Introduction

For the SDC problem, the following two problems still exist in PDF controller design:

1. Due to the lack of model knowledge, most published results only address linear precise models (see [61, 63, 64, 157, 161, 164]) that obviously cannot satisfy industrial demands. Furthermore, some nonlinear models discussed (see [58, 66, 73, 109, 183, 184, 196]) were also difficult to obtain through traditional identification approaches.

2. The constraint due to the PDF characteristics was neglected in most of the existing results (see [65, 157]). It is noted that without such a constraint, the weighting vector is irrelevant to a PDF obtained via B-spline models. In this case, the tracking objective of the output PDF cannot be achieved even if weight tracking is completed.

In order to overcome the above obstacles, in this chapter two-step NN models are applied to accomplish the SDC design problem using a constrained PI controller. After using B-spline NN approximation theory for the output PDF, we attempt to import time delay DNNs with undetermined parameters to identify the unknown nonlinear dynamic relationships. This represents a significant extension to the previous results. In the proposed method, an adaptive projection algorithm is utilized to compute the undetermined parameters related to the DNNs. Meanwhile, the PI control gains can be given based on a class of LMI algorithms. It will be shown that both the identification and tracking errors can converge to zero within a finite time using the Lyapunov stability criterion, and the state constraints can also be guaranteed simultaneously. It is noted that the proposed two-step NN approach has the independent significance in complex system control fields, and has also potential applications to other PI/PID type control problems.

## 6.2 Output PDF Model Using B-spline Neural Network

For a dynamic stochastic system, denote $u(t) \in R^m$ as the control input, $z(t) \in [a,b]$ as the stochastic output, then the probability of output $z(t)$ lying inside $[a,y]$ can be described by

$$P(a \leq z(t) < y, u(t)) = \int_a^y \gamma(\eta, u(t))d\eta \tag{6.1}$$

where $\gamma(y,u(t))$ is the PDF of the stochastic variable $y(t)$ under control input $u(t)$. Similarly to Chapter 5, square root B-spline NN are given as

$$\sqrt{\gamma(y,u(t))} = \sum_{i=1}^{n} \upsilon_i(u(t))B_i(y) \tag{6.2}$$

where $B_i(y)(i = 1,2,\cdots,n)$ are pre-specified basis functions and $\upsilon_i(t) := \upsilon_i(u(t))$, $(i = 1,2,\cdots,n)$ are the corresponding weights. Since Equation 6.2 means that $\gamma(y,u(t)) = (\sum_{i=1}^{n} \upsilon_i(u(t))B_i(y))^2 \geq 0, \forall y \in [a,b]$, it can be seen that the positiveness of $\gamma(y,u(t))$ can be automatically guaranteed. On the other hand, the output PDF should satisfy the condition $\int_a^b \gamma(y,u(t))\, dy = 1$, which means that only $n-1$ weights are independent. Therefore, square root expansions are considered as follows:

$$\gamma(y,u(t)) = (C_0(y)V(t) + \upsilon_n(t)B_n(y))^2 \tag{6.3}$$

where

$$C_0(y) = [B_1(y) \;\; B_2(y) \;\; \cdots \;\; B_{n-1}(y)],$$
$$V(t) = [\upsilon_1(t) \;\; \upsilon_2(t) \;\; \cdots \;\; \upsilon_{n-1}(t)]^T.$$

For simplicity, we denote

$$\Lambda_1 = \int_a^b C_0^T(y)C_0(y)dy, \;\; \Lambda_2 = \int_a^b C_0(y)B_n(y)dy, \;\; \Lambda_3 = \int_a^b B_n^2(y)dy. \tag{6.4}$$

To guarantee $\int_a^b \gamma(y,u(t))dy = 1$, $V^T(t)\Lambda_2\Lambda_2^T V(t) - (V^T(t)\Lambda_1 V(t) - 1)\Lambda_3 \geq 0$ should be satisfied, which is equivalent to

$$V^T(t)Q_0V(t) \leq 1 \tag{6.5}$$

where $Q_0 = \Lambda_1 - \Lambda_3^{-1}\Lambda_2^T\Lambda_2$. Under condition (6.5), $\upsilon_n(t)$ can be represented by a function of $V(t)$

$$\upsilon_n(t) = h(V(t)) = \frac{\sqrt{\Lambda_3 - V^T(t)\Lambda_0 V(t)} - \Lambda_2 V(t)}{\Lambda_3} \tag{6.6}$$

where $\Lambda_0 = \Lambda_1\Lambda_3 - \Lambda_2^T\Lambda_2$. Inequality (6.5) can be considered as a constraint on $V(t)$, which constitutes one of the major difficulties in the controller design.

## 6.3 Time Delay DNNs Identification

Once B-spline expansions have been made for the output PDFs, the next step is to find the dynamic relationships between control input and the weight vectors related to the PDFs, corresponding to a further modeling procedure.

It is well known that a DNN identifier can be employed to perform black box identification. In the following, we will provide a time delay DNN to characterize the nonlinear dynamics between $u(t)$ and $V(t)$, with a learning strategy for the model parameters. It is assumed that there exist optimal model parameters $W_1^*$, $W_2^*$ such that the unknown nonlinear dynamics between the control input $u(t)$ and the weight vectors $V(t)$ can be described by the following time delay DNN model:

$$\begin{cases} \dot{x}(t) = Ax(t) + A_1 x_\tau(t) + BW_1^* \sigma(x_\tau(t)) + BW_2^* \phi(x(t))u(t) - DF(t) \\ V(t) = Cx(t) \end{cases} \tag{6.7}$$

where $x(t) \in R^m$ is the measurable state of the unknown dynamic model and $x_\tau(t) := x(t - \tau(t))$ is the state with a time delay term. $A$ and $A_1 \in R^{m\times m}$ are stable matrices, $B$, $C$ and $D$ are known coefficient matrices with compatible dimensions. $F(t)$ represents the error term.

To formulate the required algorithm, the following assumptions are made.

**Assumption 6.1** *The time delay term $\tau(t)$ is continuous and is assumed to satisfy $\dot{\tau}(t) \le \beta < 1$, where $0 < \beta < 1$ is a known positive parameter.*

**Assumption 6.2** *Matrix $D$ satisfies the following matching conditions: there exists a known matrix $H$ such that $D = BH$.*

**Assumption 6.3** *There exists an unknown positive constant $d$ such that the error term satisfies $\|F(t)\|_1 \le d$.*

Vector functions $\sigma(.) \in R^m$ are assumed to be $m$-dimensional with elements increasing monotonically, and matrix function $\phi(.)$ is assumed to be an $m \times m$ diagonal matrix. In this context, typical presentation of the elements $\sigma_i(.)$ and $\phi_i(.)$ are as sigmoid functions, i.e.,

$$\sigma_i(x_i) = \phi_i(x_i) = \frac{a}{1 + \mathrm{e}^{-bx_i}} - c. \tag{6.8}$$

We construct the following time delay DNNs (see Figure 6.1 for details) for the system identification:

$$\begin{cases} \dot{\hat{x}}(t) = A\hat{x}(t) + A_1 \hat{x}_\tau(t) + BW_1 \sigma(\hat{x}_\tau(t)) + BW_2 \phi(\hat{x}(t))u(t) + u_f(t) \\ \hat{V}(t) = C\hat{x}(t) \end{cases} \tag{6.9}$$

where $\hat{x}(t) \in R^m$ is the state of the DNNs, the initial condition is given as $\hat{x}(t) = \varphi(t), t \in [-\tau(t), 0]$. $W_1, W_2 \in R^{m\times m}$ are the weight matrices of the DNN and $W_2$ is a diagonal matrix of the form $W_2 = \text{diag}[w_{21}, w_{22}, \cdots w_{2m}]$. It is noted that parameter $W_2^*$ is an optimally estimated value of the weight $W_2$. Therefore $W_2^*$ is a diagonal matrix. In addition, $u_f(t)$ is the compensation term and will be defined later.

Denoting $\tilde{W}_1 = W_1 - W_1^*, \tilde{W}_2 = W_1 - W_2^*$, $\tilde{\sigma} = \sigma(\hat{x}_\tau(t)) - \sigma(x_\tau(t)), \tilde{\phi} = \phi(\hat{x}(t)) - \phi(x(t))$ and the identification error as $e(t) = \hat{x}(t) - x(t)$, since $\sigma(.)$ and $\phi(.)$ are chosen as sigmoid functions, they satisfy the following Lipschitz properties

$$\tilde{\sigma}^T \tilde{\sigma} \leq e^T(t - \tau(t)) E_\sigma e(t - \tau(t)),$$

$$(\tilde{\phi} u(t))^T (\tilde{\phi} u(t)) \leq \bar{u} e^T(t) E_\phi e(t) \tag{6.10}$$

where $E_\sigma$ and $E_\phi$ are known positive definite matrices and $u(t)$ satisfies $u^T(t)u(t) \leq \bar{u}$, and $\bar{u}$ is a known constant.

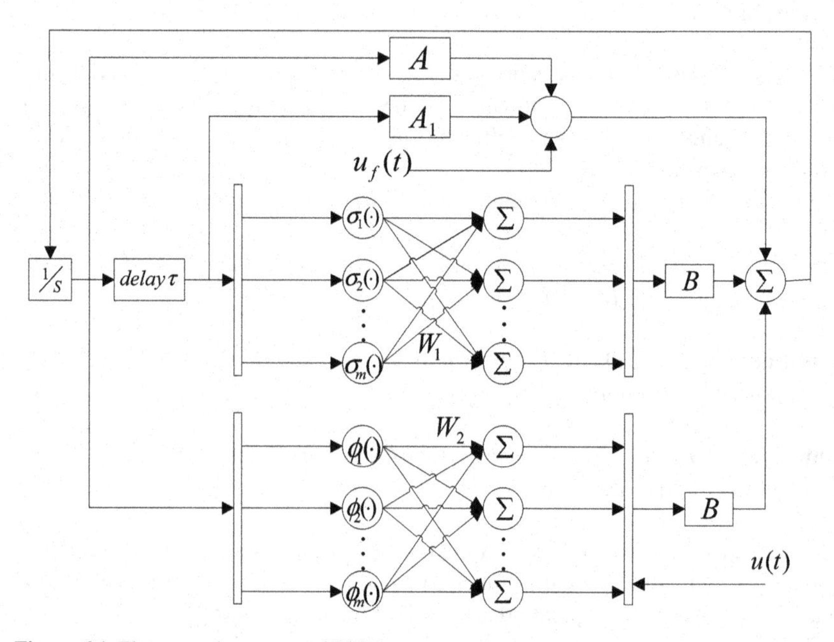


**Figure 6.1** The general structure of DNNs

From Equations 6.7 and 6.9, the identification error satisfies

$$\dot{e}(t) = Ae(t) + A_1 e_\tau(t) + B\tilde{W}_1 \sigma(\hat{x}_\tau(t)) + B\tilde{W}_2 \phi(\hat{x}(t))u(t)$$

$$+BW_1^* \tilde{\sigma} + BW_2^* \tilde{\phi} u(t) + DF(t) + u_f(t). \tag{6.11}$$

In the following, we will consider the stable learning procedure of DNN identification.

**Theorem 6.1** Consider the identification scheme (6.9), the compensation term can be given by $u_f = u_{f1} + u_{f2}$, where

$$u_{f1}(t) = -D \cdot SGN\{D^T Pe(t)\}\hat{d},$$

$$u_{f2}(t) = -\frac{1}{2}\hat{K}BB^T Pe(t), \tag{6.12}$$

and $K$ is an unknown constant and will be defined in the proof. The model parameters $W_1$ , $W_2$, $\hat{K}$ and $\hat{d}$ are updated as follows:

$$\dot{W}_1 = -\gamma_1 B^T Pe(t)\sigma^T(\hat{x}_\tau),$$

$$\dot{W}_2 = -\gamma_2 \Theta[B^T Pe(t)u^T \phi(\hat{x})],$$

$$\dot{\hat{K}} = \frac{\gamma_3}{2}\|B^T Pe(t)\|^2,$$

$$\dot{\hat{d}} = \gamma_4 \|D^T Pe(t)\|_1 \tag{6.13}$$

where $\hat{K}$ and $\hat{d}$ are estimated values of unknown constants $K$, $d$. $\gamma_i$ $(i = 1,2,3,4)$ are positive defined constants. $\Theta[.]$ represents a kind of matrix transformation that makes the matrix into a diagonal matrix. If the following LMIs

$$\Phi = \begin{bmatrix} \Psi & PA_1 \\ A_1^T P & -(1-\beta)U + E_\sigma \end{bmatrix} < 0 \tag{6.14}$$

are solvable for matrices $P > 0$, $R > 0$ and $U > 0$, where $\Psi = PA + A^T P + U + \bar{u}E_\phi + R$, then it can be obtained that $\lim_{t\to\infty} e(t) = 0$.

*Proof.* Select a Lyapunov-Krasovskii function as

$$S_1(e(t),t) = e^T(t)Pe(t) + \int_{t-\tau(t)}^{t} e^T(\alpha)Ue(\alpha)\mathrm{d}\alpha + \mathrm{tr}\{\tilde{W}_1^T \gamma_1^{-1}\tilde{W}_1\}$$
$$+\mathrm{tr}\{\tilde{W}_2^T \gamma_2^{-1}\tilde{W}_2\} + \tilde{K}^T \gamma_3^{-1}\tilde{K} + \tilde{d}^T \gamma_4^{-1}\tilde{d}. \tag{6.15}$$

According to Equation 6.11, the derivative of $S_1(e(t),t)$ can be formulated to give

$$\begin{aligned}\dot{S}_1(e(t),t) = {} & e^T(t)(\mathrm{sym}(A^T P) + U)e(t) + 2e^T(t)PA_1 e_\tau(t) - (1-\dot{\tau}(t))e_\tau^T(t)Ue_\tau(t) \\ & +2e^T(t)PDF(t) + 2e^T(t)PB\tilde{W}_1\sigma(\hat{x}_\tau) + 2e^T(t)PB\tilde{W}_2\phi(\hat{x})u(t) \\ & +2e^T(t)PBW_1^*\tilde{\sigma} + 2e^T(t)PBW_2^*\tilde{\phi}u(t) + 2e^T(t)Pu_{f1}(t) \\ & +2e^T(t)Pu_{f2}(t) + 2\dot{\tilde{K}}^T \gamma_3^{-1}\tilde{K} + 2\dot{\tilde{d}}^T \gamma_4^{-1}\tilde{d} \\ & +2\mathrm{tr}\{\dot{\tilde{W}}_1^T \gamma_1^{-1}\tilde{W}_1\} + 2\mathrm{tr}\{\dot{\tilde{W}}_2^T \gamma_2^{-1}\tilde{W}_2\}.\end{aligned}$$

Since $e^T PBW_1^*\tilde{\sigma}$ and $e^T PBW_2^*\tilde{\phi}u(t)$ are scalars, it can be obtained that

$$2e^T PBW_1^* \tilde{\sigma} \le e^T PBW_1^* W_1^{*T} B^T Pe + \tilde{\sigma}^T \tilde{\sigma},$$

$$2e^T PBW_2^* \tilde{\phi} u(t) \le e^T PBW_2^* W_2^{*T} B^T Pe + (\tilde{\phi} u(t))^T \tilde{\phi} u(t). \tag{6.16}$$

Denote $W_1^* W_1^{*T} = \bar{W}_1^*$, $W_2^* W_2^{*T} = \bar{W}_2^*$, $K = \|\bar{W}_1^* + \bar{W}_2^*\|$, according to inequality (6.10), we have

$$\begin{aligned}
\dot{S}_1(t) &\le e^T(t)[\mathrm{sym}(A^T P) + U + \bar{u}E_\phi]e(t) + 2e^T(t)PA_1 e_\tau(t) + 2e^T(t)PB\tilde{W}_1\sigma(\hat{x}_\tau) \\
&\quad - e_\tau^T(t)[(1-\beta)U - E_\sigma]e_\tau(t) + 2e^T(t)PB\tilde{W}_1\sigma(\hat{x}_\tau) + 2e^T(t)PB\tilde{W}_2\phi(\hat{x})u(t) \\
&\quad + 2\mathrm{tr}\{\dot{\tilde{W}}_1^T \gamma_1^{-1}\tilde{W}_1\} + 2\mathrm{tr}\{\dot{\tilde{W}}_2^T \gamma_2^{-1}\tilde{W}_2\} + 2e^T(t)Pu_{f1}(t) + 2e^T(t)Pu_{f2}(t) \\
&\quad + 2e^T(t)PDF(t) + e^T(t)PB(\bar{W}_1^* + \bar{W}_2^*)B^T Pe(t) + 2\dot{\tilde{K}}^T \gamma_3^{-1}\tilde{K} + 2\dot{\tilde{d}}^T \gamma_4^{-1}\tilde{d} \\
&\le e^T(t)[\mathrm{sym}(A^T P) + U + \bar{u}E_\phi + R]e(t) - e^T(t)Re(t) + 2e^T(t)PA_1 e_\tau(t) \\
&\quad - e_\tau^T(t)[(1-\beta)U - E_\sigma]e_\tau(t) + 2e^T(t)PB\tilde{W}_1\sigma(\hat{x}_\tau) + 2e^T(t)PB\tilde{W}_2\phi(\hat{x})u(t) \\
&\quad + 2\mathrm{tr}\{\dot{\tilde{W}}_1^T \gamma_1^{-1}\tilde{W}_1\} + 2\mathrm{tr}\{\dot{\tilde{W}}_2^T \gamma_2^{-1}\tilde{W}_2\} - 2e^T(t)PD \cdot SGN\{D^T Pe\}\hat{d} \\
&\quad + 2\dot{\tilde{d}}^T \gamma_4^{-1}\tilde{d} + 2e^T(t)PA_1 e_\tau(t) + 2\dot{\tilde{K}}^T \gamma_3^{-1}\tilde{K} - \hat{K}\|B^T Pe(t)\|^2 \\
&\quad + K\|B^T Pe(t)\|^2.
\end{aligned} \tag{6.17}$$

Since $\dot{\tilde{W}}_1 = \dot{W}_1, \dot{\tilde{W}}_2 = \dot{W}_2, \dot{\tilde{K}} = \dot{\hat{K}}, \dot{\tilde{d}} = \dot{\hat{d}}$, if we use the updating laws (6.13) and LMI (6.14), we have

$$\dot{S}_1(t) \le \theta^T(t)\Phi\theta(t) - e^T(t)Re(t) \le -e^T(t)Re(t) \le 0 \tag{6.18}$$

where $\theta(t) = [e^T(t), e^T(t-\tau(t))]^T$. Hence, $S_1(t) \in L_\infty$, which implies that $e(t) \in L_\infty$. From Equation 6.11, it can be seen that $\dot{e} \in L_\infty$. Furthermore, by integrating both sides of inequality (6.18), it can be seen that

$$\int_0^\infty \|e(t)\|_R^2 \mathrm{d}t \le S_1(0) - S_1(\infty) < \infty. \tag{6.19}$$

As a result $e(t) \in L_2 \bigcap L_\infty$. Using the well-known Barbalat Lemma [39], we have $\lim_{t\to\infty} e(t) = 0$. Furthermore, it can be obtained that $\lim_{t\to\infty} \hat{V}(t) = V(t)$. □

*Remark* 6.1 It is well known that adaptive compensation techniques have been widely applied to sliding mode control, robot control, dead zero control and so on. Compared with other results related to DNNs, we consider the identification error $F(t)$ and construct the compensation term $u_{f1}(t)$ with saturation function to guarantee the error $e(t)$ converging to zero. On the other hand, in [189], the optimal weight values $W_1^*$, $W_2^*$ exist and are bounded (i.e., $W_1^* W_1^{*T} \le \bar{W}_1$ and $W_2^* W_2^{*T} \le \bar{W}_2$). However, their boundaries $\bar{W}_1$ and $\bar{W}_2$ are unknown and cannot be utilized in many practical processes. In the chapter, the compensation term $u_{f2}(t)$ is designed to compensate the influence of the unknown boundary and guarantee the identification error converging to zero.

*Remark* 6.2 Time delays are frequently encountered in many practical engineering systems. It is now well known that time delay is one of the main causes of instability and poor performance of control systems. Similarly to [126, 131, 191], we consider the time delay term in DNNs. In our opinion, compared with applying the DNN to identify a nonlinear systems with no time delay term, it is a reasonable generalization to use the DNN including time delay term to identify a nonlinear system with time delay term. For example, considering the following time delay nonlinear system

$$\dot{x}(t) = f(x(t), x(t-\tau(t)), u(t)), \quad y(t) = h(x(t), u(t)). \tag{6.20}$$

Obviously, only DNNs with time delay can exactly identify the above nonlinear system (see [126, 131, 191]).

## 6.4 Constrained PI Tracking Control for Weight Vector

In this section, we investigate the tracking controller design for DNNs (6.9). Corresponding to Equation 6.3, a desired (known) PDF can be described by

$$g(y) = (C_0(y)V_g + h(V_g)B_n(y))^2 \tag{6.21}$$

where $V_g$ is the desired weight vector corresponding to $B_i(y)$. The tracking objective is to find $u(t)$ such that $\gamma(y, u(t))$ can follow $g(y)$. The error between the output PDF and the target PDF is formulated by $\Delta_e = \sqrt{g(y)} - \sqrt{\gamma(y, u(t))}$, i.e.,

$$\Delta_e = C_0 V_e + [h(V_g) - h(V(t))]B_n(y) \tag{6.22}$$

where $V_e(t) = V_g - V(t)$. Due to the continuity of $h(V(t))$, $\Delta_e \longrightarrow 0$ holds as long as $V_e \longrightarrow 0$. At this stage, the PDF control problem can be formulated into the tracking problem for the above nonlinear weight systems, and the control objective is to find $u(t)$ such that the tracking performance, state constraints and stability can all be guaranteed simultaneously.

Based on DNN model (6.9), we introduce a new state variable

$$\bar{x}(t) := [\hat{x}^T(t), \int_0^t (V_g - \hat{V}(\alpha))^T \mathrm{d}\alpha]^T. \tag{6.23}$$

Then the DNNs can be transformed into an equivalent nonlinear form:

$$\dot{\bar{x}}(t) = \bar{A}\bar{x}(t) + \bar{A}_1\bar{x}_\tau(t) + \bar{B}W_1\sigma(\hat{x}_\tau(t)) + \bar{B}W_2\phi(\hat{x}(t))u(t) + \bar{B}\bar{u}_f(t) + \bar{I}V_g \tag{6.24}$$

where

$$\bar{u}_{f1}(t) = -H \cdot SGN\{D^T Pe(t)\}\hat{d},$$

$$\bar{u}_{f2}(t) = -\frac{1}{2}\hat{K}B^T Pe(t),$$

$$\bar{A} = \begin{bmatrix} A & 0 \\ -C & 0 \end{bmatrix}, \ \bar{A}_1 = \begin{bmatrix} A_1 & 0 \\ 0 & 0 \end{bmatrix}, \ \bar{B} = \begin{bmatrix} B \\ 0 \end{bmatrix}, \ \bar{I} = \begin{bmatrix} 0 \\ I \end{bmatrix}.$$

To solve the tracking problem for weight vectors, a control law consisting of the PI control strategy and an adaptive compensator is designed as follows:

$$u(t) = [W_2\phi(\hat{x}(t))]^{-1}[K_P\hat{x}(t) + K_I\int_0^t (V_g - \hat{V}(\alpha))^T \mathrm{d}\alpha - W_1\sigma(\hat{x}_\tau(t)) - \bar{u}_f(t)] \tag{6.25}$$

where $K = [K_P, K_I]$ are the PI controller gains to be determined.

Substituting $u(t)$ into Equation 6.24, the corresponding nonlinear dynamic model can be described by

$$\dot{\bar{x}}(t) = (\bar{A} + \bar{B}K)\bar{x}(t) + \bar{A}_1\bar{x}_\tau(t) + \bar{I}V_g. \tag{6.26}$$

At this stage, the following theorem can be obtained.

**Theorem 6.2** Consider the dynamic nonlinear model (6.24), when control law (6.25) and adaptive law (6.13) are applied, and for the known parameters $\mu$, $\eta$ and known matrix $T$, suppose that there exist $Q = P_1^{-1} > 0$, $\bar{S} > 0$ and $G$ such that the following LMIs

$$\begin{bmatrix} \Upsilon & \bar{A}_1 Q & \bar{I} \\ Q\bar{A}_1^T & -(1-\beta)\bar{S} & 0 \\ \bar{I}^T & 0 & -\mu^2 I \end{bmatrix} < 0 \tag{6.27}$$

where $\Upsilon = \mathrm{sym}(Q\bar{A}^T) + \mathrm{sym}(\bar{B}G) + \bar{S} + \eta^2 Q$.

$$\begin{bmatrix} Q & Q\bar{T} \\ \bar{T}Q & \eta^2\mu^{-2}r^{-1} \end{bmatrix} \geq 0, \tag{6.28}$$

$$\begin{bmatrix} Q & Q\bar{T} \\ \bar{T}Q & I \end{bmatrix} \geq 0, \ \begin{bmatrix} 1 & \bar{x}_m^T \\ \bar{x}_m & Q \end{bmatrix} \geq 0 \tag{6.29}$$

are solvable, then the closed-loop system (6.26) is stable and satisfies $\lim_{t\to\infty}\hat{V}(t) = V_g$, $\hat{V}^T(t)Q_0\hat{V}(t) \leq 1$ simultaneously. In this case, the control gain is given by $K = GQ^{-1}$, and $\bar{S} = QSQ$.

*Proof.* Select a Lyapunov-Krasovskii function

$$S_2(\bar{x}(t),t) = \bar{x}^T(t)P_1\bar{x}(t) + \int_{t-\tau(t)}^t \bar{x}^T(\alpha)S\bar{x}(\alpha)\mathrm{d}\alpha. \tag{6.30}$$

According to Equation 6.26, the derivative of $S_2(\bar{x}(t),t)$ can be formulated to give

$$\dot{S}_2(\bar{x}(t),t) = \bar{x}^T(t)(\mathrm{sym}(P_1\bar{A} + P_1\bar{B}K)) + S)\bar{x}(t) + 2\bar{x}^T(t)P_1\bar{A}_1\bar{x}_\tau(t)$$

$$-(1-\dot{\tau}(t))\bar{x}_\tau^T(t)S\bar{x}_\tau(t)+2\bar{x}^T(t)P_1\bar{I}V_g$$

$$\leq \bar{x}^T(t)(\mathrm{sym}(P_1\bar{A}+P_1\bar{B}K)+S+\eta^2P_1+\frac{1}{\mu^2}P_1\bar{I}\bar{I}^TP_1)\bar{x}(t)$$

$$+2\bar{x}^T(t)P_1\bar{A}_1\bar{x}_\tau(t)-(1-\beta)\bar{x}_\tau^T(t)S\bar{x}_\tau(t)+\mu^2\|V_g\|^2.$$

Therefore we have

$$\dot{S}_2(\bar{x}(t),t)\leq\bar{\theta}^T(t)\Phi_1\bar{\theta}(t)+\mu^2r \tag{6.31}$$

where $\bar{\theta}(t)=[\bar{x}^T(t),\bar{x}_\tau^T(t)]^T$. Since $V_g$ can be a known vector, we denote $r:=\|V_g\|^2$ and

$$\Phi_1=\begin{bmatrix}\Upsilon_1 & P_1\bar{A}_1\\ \bar{A}_1^TP_1 & -(1-\beta)S\end{bmatrix}$$

where $\Upsilon_1=\mathrm{sym}(\bar{A}^TP_1)+\mathrm{sym}(K^T\bar{B}^TP_1)+S+\frac{1}{\mu^2}P_1\bar{I}\bar{I}^TP_1+\eta^2P_1$. By pre-multiplying $\mathrm{diag}\{P_1^{-1},P_1^{-1}\}$ and post-multiplying $\mathrm{diag}\{P_1^{-1},P_1^{-1}\}$ on both sides of $\Phi_1$, based on the Schur complement formula and defining $Q=P_1^{-1}$, it can be shown that $\Phi_1<0$ is equivalent to the following LMI:

$$\begin{bmatrix}\Upsilon_2 & \bar{A}_1 & \bar{I}\\ \bar{A}_1^T & -(1-\beta)QSQ & 0\\ \bar{I}^T & 0 & -\mu^2I\end{bmatrix}<0 \tag{6.32}$$

where $\Upsilon_2=\mathrm{sym}(Q\bar{A}^T)+\mathrm{sym}(\bar{B}KQ)+QSQ+\eta^2Q$. Defining $K=GP_1$, $\bar{S}=QSQ$, then LMI (6.32) is equivalent to LMI (6.27). Therefore it can be obtained that

$$\dot{S}_2(\bar{x}(t),t)\leq-\eta^2\bar{x}^T(t)P_1\bar{x}(t)+\mu^2r. \tag{6.33}$$

Thus, $\frac{\mathrm{d}s_2(\bar{x}(t),t)}{\mathrm{d}t}<0$, only if $\bar{x}^T(t)P_1\bar{x}(t)>\eta^{-2}\mu^2r$ holds. As such, for any $\bar{x}(t)$, it can be verified that

$$\bar{x}^T(t)P_1\bar{x}(t)\leq\max\{\bar{x}_m^TP_1\bar{x}_m,\eta^{-2}\mu^2r\},$$

$$\|\bar{x}_m\|=\sup_{-\tau(t)\leq t\leq 0}\|\bar{x}(t)\| \tag{6.34}$$

for all $t\in[-\tau(t),\infty)$, which also implies that the system is stable.

Consider two trajectories $\bar{x}_1(t)$ and $\bar{x}_2(t)$ of system (6.26) corresponding to the determined input $V_g$. Then the dynamics of $\sigma(t)=\bar{x}_1(t)-\bar{x}_2(t)$ can be described by

$$\dot{\sigma}(t)=(\bar{A}+\bar{B}K)\sigma(t)+\bar{A}_1\sigma_\tau(t). \tag{6.35}$$

Similarly to Equation 6.30, a Lyapunov-Krasovskii function can be constructed as

$$S_2(\sigma(t),t)=\sigma^T(t)P_1\sigma(t)+\int_{t-\tau(t)}^{t}\sigma^T(\alpha)S\sigma(\alpha)\mathrm{d}\alpha. \tag{6.36}$$

From inequality (6.27) it can be seen that

$$\dot{S}_2(\sigma(t),t) \leq -\eta^2\lambda_{\min}(P_1)\|\sigma(t)\|^2$$

where $\lambda_{\min}(P_1)$ is the minimum eigenvalue of $P_1$. Thus, Equation 6.26 is asymptotically stable with respect to $\sigma = 0$, and there exists only one equilibrium $\bar{x}^*$. Consequently, $\lim_{t\to\infty}\frac{\mathrm{d}}{\mathrm{d}t}\int_0^t (V_g - \hat{V}(\alpha))^T \mathrm{d}\alpha = 0$ holds, which shows that $\lim_{t\to\infty}\hat{V}(t) = V_g$.

It has been shown that due to the property of a PDF, the observed weight vectors have to satisfy $\hat{V}^T(t)Q_0\hat{V}(t) \leq 1$, which can be reduced to $\bar{x}^T(t)T\bar{x}(t) \leq 1$, where $T := \mathrm{diag}\{C^T Q_0 C, 0\}$. Based on the property of a non-negative definite matrix, $T$ can be divided into $T = \bar{T}^2$, where $\bar{T} \geq 0$. Similarly to the above proof, inequality (6.34) still holds. This means that $\bar{x}^T(t)P_1\bar{x}(t) \leq \bar{x}_m^T P_1 \bar{x}_m$ or $\bar{x}^T(t)P_1\bar{x}(t) \leq \eta^{-2}\lambda^2 r$. Combining with LMI (6.29), $T \leq P_1$ and $\bar{x}_m^T P_1 \bar{x}_m \leq 1$ can be satisfied, and it can be obtained that

$$\bar{x}^T(t)T\bar{x}(t) \leq \bar{x}^T(t)P_1\bar{x}(t) \leq \bar{x}_m^T P_1 \bar{x}_m \leq 1. \tag{6.37}$$

On the other hand, from LMI (6.28), $T \leq \eta^2\lambda^{-2}r^{-1}P_1$, it can be seen that

$$\bar{x}^T(t)T\bar{x}(t) \leq \eta^2(\lambda^2 r)^{-1}\bar{x}^T(t)P_1\bar{x}(t) \leq 1. \tag{6.38}$$

Based upon inequalities (6.37) and (6.38), it can be seen that constraint condition $\hat{V}^T(t)Q_0\hat{V}(t) \leq 1$ can be satisfied. □

*Remark* 6.3 Note that when the LMIs (6.14, 6.27, 6.28, 6.29) are solvable to $P$ and $P_1$ respectively, then the control input $u(t)$ satisfies Equation 6.25, the adaptive law accords with Equation 6.13 and for the designed $\bar{u}$, we have $u^T(t)u(t) < \bar{u}$. This means that both the tracking error and the identification error can be made to converge to zero at the same time. Compared with other results about output PDF tracking control, such as [63, 157], this chapter cannot only accomplish the dynamic tracking control problem through the designed control input, but also identify the dynamic trajectory of the weight vectors related to the output PDFs through the proposed DNN identifier.

In order to ensure the existence of $[W_2\phi(\hat{x}(t))]^{-1}$, we need to establish that $w_{2i} \neq 0,\ \ i = 1,\cdots m$. Furthermore, from Equation 6.13 it can be seen that weight matrices $W_1$, $W_2$ cannot be guaranteed within reasonable bounds. Hence $W_1$ and $W_2$ are redefined through the use of projection algorithm. In particular, the standard adaptive laws are modified to read

$$\dot{W}_1 = \begin{cases} -\gamma_1 B^T Pe(t)\sigma^T(\hat{x}_\tau) & when\ \ \|W_1\| < M_1 or\ \ \|W_1\| = M_1 \\ & and\ \ \mathrm{tr}\{\sigma(\hat{x}_\tau)e^T(t)PBW_1\} \geq 0 \\ -\gamma_1 B^T Pe(t)\sigma^T(\hat{x}_\tau) + \gamma_1 \mathrm{tr}\{\sigma(\hat{x}_\tau)e^T(t)PBW_1\}\frac{W_1}{\|W_1\|^2} & \\ \quad when\ \ \|W_1\| = M_1 and\ \ \mathrm{tr}\{\sigma(\hat{x}_\tau)e^T(t)PBW_1\} < 0 & \end{cases}. \tag{6.39}$$

When $w_{2i} = \varepsilon$, we adopt

$$\dot{w}_{2i} = \begin{cases} -\gamma_2 u_i\phi_i(\hat{x})B_i^T Pe(t) & when\ \ u_i\phi_i(\hat{x})B_i^T Pe(t) < 0 \\ 0 & when\ \ u_i\phi_i(\hat{x})B_i^T Pe(t) \geq 0 \end{cases} \tag{6.40}$$

where $u_i$ is the ith element of $u(t)$ and $B_i$ is the ith column of $B$. Otherwise (i.e., when $w_{2i} \neq \varepsilon$), we have

$$\dot{W}_2 = \begin{cases} -\gamma_2 \Theta[B^T Pe(t)u^T(t)\phi(\hat{x})] \ \ when \ \ \|W_2\| < M_2 \ \ or \ \ \|W_2\| = M_2 \\ \qquad\qquad and \ \ \mathrm{tr}\{\phi^T(\hat{x})u(t)e^T(t)PBW_2\} \geq 0 \\ -\gamma_2 \Theta[B^T Pe(t)u^T(t)\phi(\hat{x})] + \gamma_2 \mathrm{tr}\{\phi^T(\hat{x})u(t)e^T(t)PBW_2\} \frac{W_2}{\|W_2\|^2} \\ \qquad when \ \ \|W_2\| = M_2 \ \ and \ \ \mathrm{tr}\{\phi^T(\hat{x})u(t)e^T(t)PBW_2\} < 0. \end{cases} \tag{6.41}$$

*Remark* 6.4 The modified adaptive laws (6.39)-(6.41) can satisfy our requirements better than those of [170, 171]. Moreover, (6.39) and (6.41) can guarantee the uniform boundness for weights $W_1$ and $W_2$. Define $\Xi_1$ and $\Xi_2$ as constraint sets for $W_1$ and $W_2$, respectively, that is $\Xi_i = \{W_i : \ \mathrm{tr}(W_i W_i^T) \leq M_i, M_i > 0\}, \ \ i = 1,2$. In Equation 4.41, $\Theta[.]$ is applied because $W_2$ is a diagonal matrix. The initial value of $w_{2i}$ is chosen to be larger than $\varepsilon$ so that $w_{2i} \geq \varepsilon$ can be ensured through Equation 6.40.

*Remark* 6.5 From the expression (6.25) for input $u(t)$, the weights $W_1, W_2$, the states $x(t)$, $\hat{x}(t)$ and the functions $\phi(\hat{x})$, $\sigma(\hat{x}_\tau)$ can be proved bounded, so we can prove that the input is also bounded and the assumption $u^T(t)u(t) < \bar{u}$ is reasonable. Of course, for online identification, it is a limitation and may cause the recurrent proof process. However, in much academic research into NN control or adaptive control, the recurrent proof process exists and is difficult to avoid. We will consider this problem carefully in the future and try to solve it.

## 6.5 Illustrative Examples

### 6.5.1 Example 1

In many practical processes, such as particle distribution control problems, the shapes of measured output PDF normally have two or three peaks. This means that one can consider the PDF shape control to follow a curve of three peaks. As such, the target PDF should also be a curve with three peaks in line with the specification of the product quality.

In this example, we suppose that the output PDFs can be approximated using the square root B-spline models described by Equation 6.3 with $n = 3$, $y \in [0, 1.5]$ and $i = 1,2,3$ with the following basis functions:

$$B_i(y) = \begin{cases} |\sin 2\pi y| & y \in [0.5(i-1); 0.5i] \\ 0 & y \in [0.5(j-1); 0.5j] \ \ i \neq j. \end{cases} \tag{6.42}$$

From the notation in Equation 6.4, it can be seen that $\Lambda_1 = \mathrm{diag}\{0.25, 0.25\}, \Lambda_2 = [0,0], \Lambda_3 = 0.25$. The desired PDF $g(y)$ is described by Equation 6.21 with $V_g = [0.9, 1.2]^T$.

In this simulation, we assume that the unknown dynamic model is designed as

$$\dot{x}(t) = f(x(t), x_\tau(t)) + g(x(t))u(t) + d_1(t) \tag{6.43}$$

where

$$f(x) = \begin{bmatrix} \sin(-3x_1 - 2x_2) - x_{1\tau} \\ \sin x_2 - x_{2\tau} \\ -2x_2 + x_3 - x_{3\tau} \end{bmatrix}, \quad d_1(t) = \begin{bmatrix} 0.2 \\ 0.2 \\ -0.2 \end{bmatrix},$$

$$g(x) = \begin{bmatrix} 2.6 & 0 & 0 \\ 0 & 2.6 & 0 \\ 0 & 0 & 4 \end{bmatrix}, \quad x(0) = \begin{bmatrix} 0.1 \\ 0.1 \\ 0.2 \end{bmatrix}.$$

For DNN model (6.9), we select

$$\sigma(x_i) = \phi(x_i) = \frac{2}{1 + \mathrm{e}^{-0.5x_i}} + 0.5 \quad i = 1,2,3, \quad \hat{x}(t) = [0.1, 0, 0.4]^T \quad t \in [-\tau(t), 0],$$

$$A = \begin{bmatrix} -2 & 0 & -2 \\ 0 & -2 & 0 \\ 2 & 0 & -2 \end{bmatrix}, \quad B = \begin{bmatrix} 1 & 0 & 0 \\ 0 & 1 & 0 \\ 0 & 0 & 1 \end{bmatrix}, \quad C = \begin{bmatrix} -1 & 0 & 1 \\ 0 & 1 & 1 \end{bmatrix}, \quad A_1 = \begin{bmatrix} -1 & 0 & 0 \\ 0 & -1 & 0 \\ 0 & 0 & -1 \end{bmatrix}.$$

For adaptive projection algorithms, we choose

$$M_1 = M_2 = 8, \quad \varepsilon = 0.01, \quad \gamma_i = 1 \quad i = 1,2,3,4,$$

$$W_{10} = \begin{bmatrix} 1 & 0 & 0 \\ 0 & 1 & 0 \\ 0 & 0 & 1 \end{bmatrix}, \quad W_{20} = \begin{bmatrix} 2 & 0 & 0 \\ 0 & 2 & 0 \\ 0 & 0 & 2 \end{bmatrix}.$$

By defining $\bar{u} = 8$, $\beta = 0.8$, $\mu = -2$, $\eta = 2$ and solving LMIs (6.14, 6.27, 6.28, 6.29), it can be obtained that

$$P = \begin{bmatrix} 0.4585 & 0 & 0 \\ 0 & 0.4585 & 0 \\ 0 & 0 & 0.4585 \end{bmatrix},$$

$$Q = \begin{bmatrix} 4.9929 & 3.2831 & -1.4026 & -2.6468 & 0.3221 \\ 3.2831 & 6.8594 & -2.5106 & -2.4513 & 1.8585 \\ -1.4026 & -2.5106 & 7.4443 & 3.3798 & 2.4565 \\ -2.6468 & -2.4513 & 3.3798 & 3.5060 & 0.9569 \\ 0.3221 & 1.8585 & 2.4565 & 0.9569 & 2.5924 \end{bmatrix},$$

$$K_P = \begin{bmatrix} -6.7779 & 18.2598 & 10.8267 \\ -12.9210 & -3.9561 & -2.0216 \\ 5.1086 & -2.7023 & -5.3602 \end{bmatrix}, \quad K_I = \begin{bmatrix} 3.0520 & -21.0130 \\ -16.0323 & 13.9521 \\ 9.9798 & 4.3526 \end{bmatrix}.$$

With control input (6.25) and adaptive projection algorithms applied, the responses of the identified nonlinear system and the states of DNNs are shown in Figure 6.2-6.4. Figure 6.5 shows the trajectory of the identified weight vectors. The

3D mesh plot of the output PDFs are shown in Figure 6.6. Figure 6.2-6.6 demonstrate that satisfactory tracking performance, identification capability and robustness have been achieved.

On the other hand, in order to show the extensive applicability of the proposed method, we consider the following system:

$$\sqrt{\gamma, u(t)} = B_1(y)w_1 + B_2(y)w_1 + B_3(y)w_1$$

where $a$ and $b$ are set to -3 and +2, respectively, and

$$B_1(y) = [y^2+6y+9]I_1 + [-y^2-3y-1]I_2 + [y^2]I_3,$$

$$B_2(y) = [y^2+4y+4]I_2 + [-y^2-y+1]I_3 + [y^2-2y+1]I_4,$$

$$B_3(y) = [y^2+2y+1]I_3 + [-y^2+y+1]I_4 + [y^2-4y+4]I_5.$$

For this system, $I_i$ are the interval functions, defined as follows:

$$I_i(y) = \begin{cases} 1 & y \in [\lambda_i; \lambda_{i+1}) \\ 0 & elsewhere \end{cases} \tag{6.44}$$

where $\lambda_i = i-4, \;\; i = 1,2,3,4,5$. From the notation in Equation 6.4, it can be seen that

$$\Lambda_1 = \begin{bmatrix} 1.7667 \; 0.7667 \\ 0.7667 \; 1.7667 \end{bmatrix}, \;\; \Lambda_2 = [0.0333, 0.07667], \;\; \Lambda_3 = 1.7667.$$

The desired PDF $g(y)$ is described by Equation 6.21 with $V_g = [0.3, 0.3]^T$. Other parameters are the same as the previous design. Based on new B-spline models, the 3D mesh plot of output PDFs are showed in Figure 6.7. Figure 6.6 and 6.7 demonstrate that satisfactory tracking performance can be obtained for different B-spline models and target PDFs.

### 6.5.2 Example 2

A typical example of a paper machine is shown in Figure 5.8. In the head box, fiber, fillers and other chemical additives are mixed. This mixture generally consists of solids and water. When this mixture is injected onto the moving wire table, some water is drained through the wire nets into a white water pit underneath the wire table. This drainage process is continuous and the water in the pit also contains some solids in the form of either flocculation, or particles with a size distribution. As such, in order to control the efficiency of raw material usage, the total solids in the drained water need to be controlled and minimized.

As discussed in the previous chapter, with reference to several laboratory tests on the flocculation size distribution in the white water system, 100 samples have been formulated and used to evaluate 100 different PDFs. The distribution of flocculation sizes can be approximated by a truncated $\Gamma$-distribution under certain assumptions.

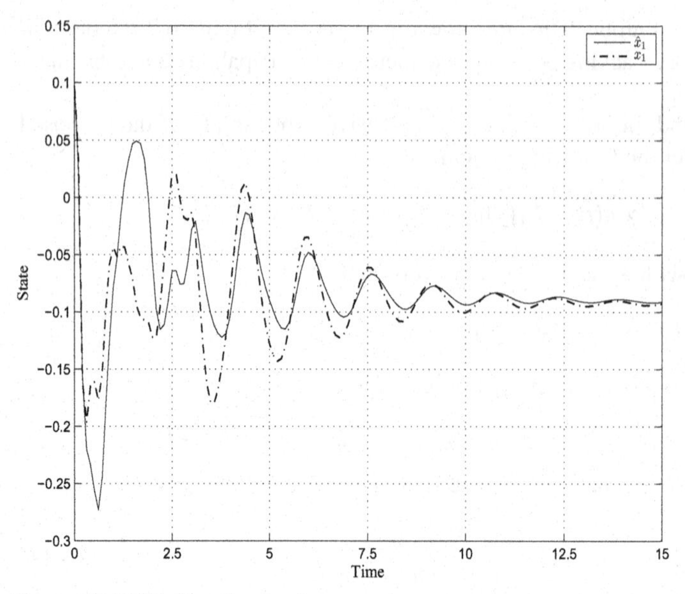


**Figure 6.2** DNNs identification for $x_1$

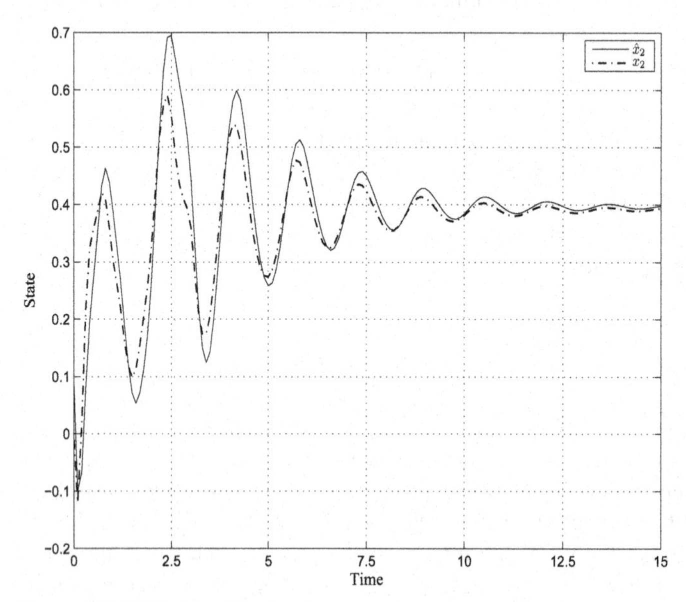


**Figure 6.3** DNNs identification for $x_2$

Since this distribution involves an exponential function, it is possible to use a set of exponential type spline functions to approximate the actual flocculation size distribution, given by the following:

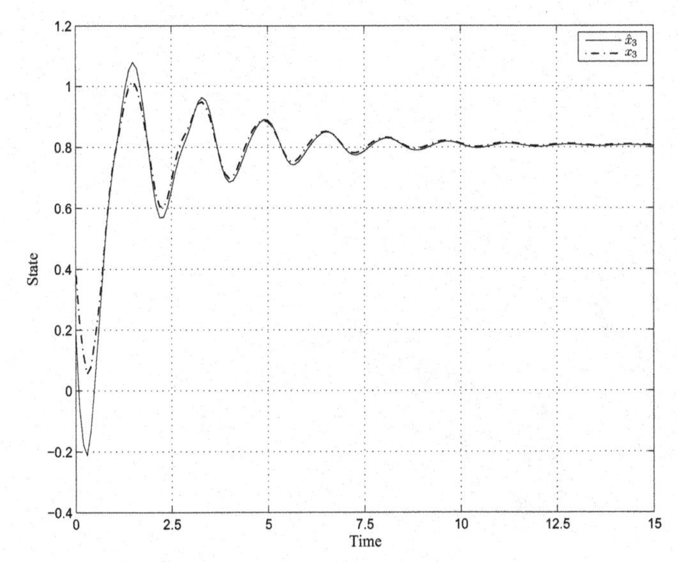


**Figure 6.4** DNNs identification for $x_3$

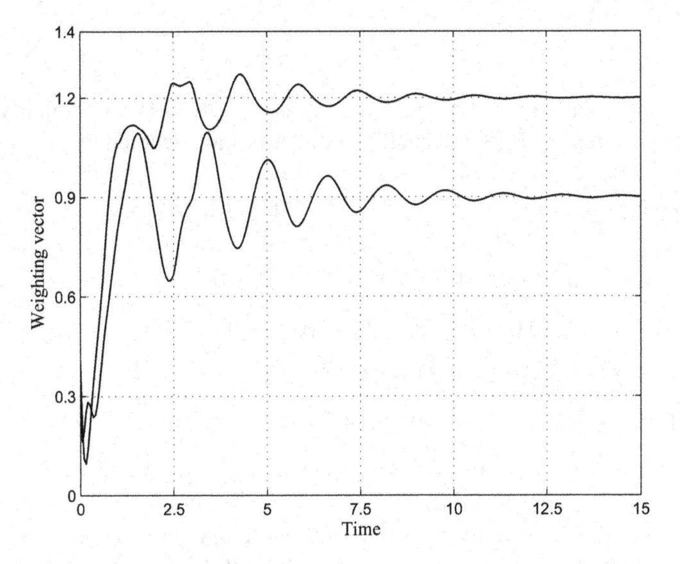


**Figure 6.5** Dynamic weighting vectors $\hat{V}(t)$

$$B_i(y) = \exp(-(y-y_i)^2\sigma_i^{-2}) \quad (i = 1, 2, \cdots, 5)$$

where $V(t) = [v_1, v_2, v_3, v_4, v_5]^T$, $y \in [0, 0.05], y_i = 0.003 + 0.006(i-1), \sigma_i = 0.03$ $(i = 1, 2, \cdots, 5)$. The desired PDF $g(y)$ is assumed to be described by Equation 6.21 with $V_g = [0.5, 0.6, 0.7, 0.8]^T$.

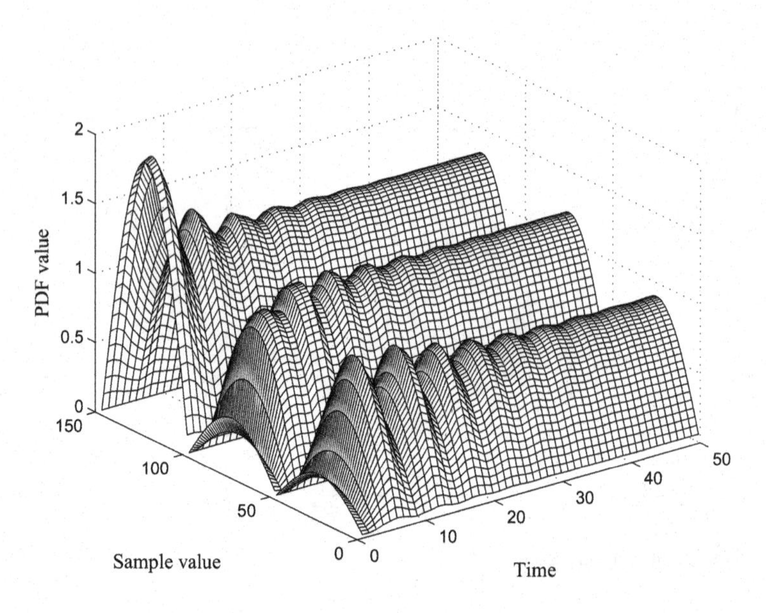


**Figure 6.6** 3D mesh plot of the output PDFs

The dynamic model relating $u(t)$ and $x(t)$ is described by DNNs (6.9). Similarly to the modeling procedures of [161], the coefficient matrices are denoted by

$$\sigma(x_i) = \phi(x_i) = \frac{2}{1+e^{-0.5x_i}} + 0.5 \quad i = 1,2,3,4,5,$$

$$A = \mathrm{diag}\{-0.083, -0.083, -0.083, -0.083, -0.083\},$$

$$A_1 = \mathrm{diag}\{-0.02, -0.02, -0.02, -0.02, -0.02\},$$

$$B = \mathrm{diag}\{0.78, 0.78, 0.78, 0.78, 0.78\},$$

$$C = [1\ 0\ 1\ 0\ 1, 0\ 1\ -1\ 0\ 0, 1\ 0\ 0\ 0\ 1, 0\ -1\ 1\ 1\ 1].$$

The unknown dynamic model, the adaptive laws and other parameters are the same as Example 1. By solving LMIs (6.14, 6.27-6.29), we can obtain values for $K_p, K_I, P$ and $Q$ (due to the large dimensions of these matrices, their element values are omitted in this chapter). In order to show the influence of the time delay on the control system, we consider three different time delays: 0.2 s, 0.3 s and 0.5 s. The responses of weight vectors with different time delays are shown in Figure 6.8-6.10. It can be concluded that the time delay is one of the main causes of poor closed performance. When the time delay increases, the convergence speed is reduced and the tracking performance gets worse. Finally, the 3D mesh plots of the output PDFs for $\tau(t) = 0.3$ s are shown in Figure 6.11.

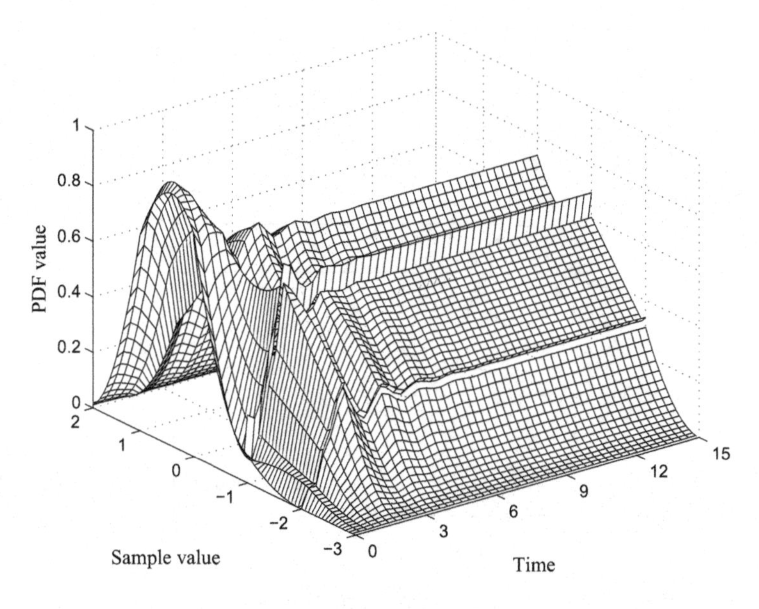


**Figure 6.7** 3D mesh plot of the output PDFs

## 6.6 Conclusions

In this chapter, two-step NN are employed to solve the tracking control problem for the conditional output PDF of non-Gaussian processes using a generalized PI controller. After the B-spline approximation to the measured output PDFs, DNNs can be used to describe the complex nonlinear relationships between the control input and the weights related to output PDF. Using LMI methods and adaptive projection algorithms, a constrained PI tracking controller can be obtained such that the closed-loop tracking performance, system identification and state constraints are guaranteed simultaneously. Two simulations are given to demonstrate the effectiveness of the proposed approach.

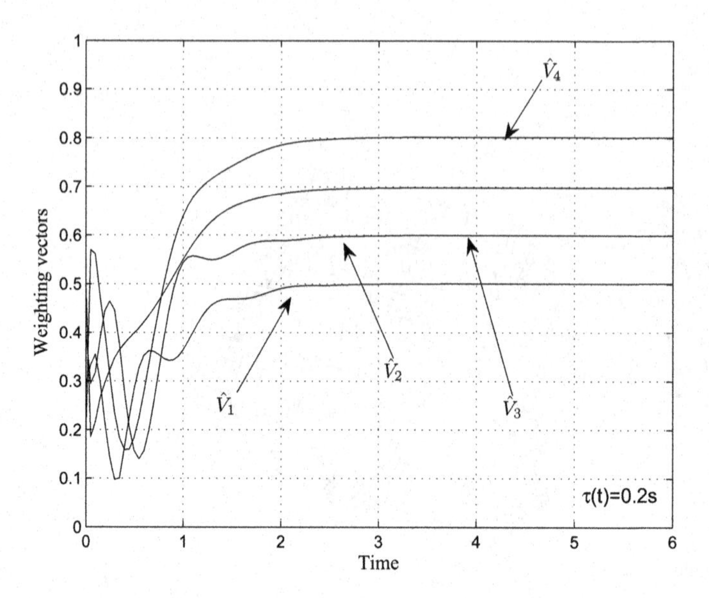


**Figure 6.8** Responses of dynamic weighting vectors for $\tau(t) = 0.2$ s

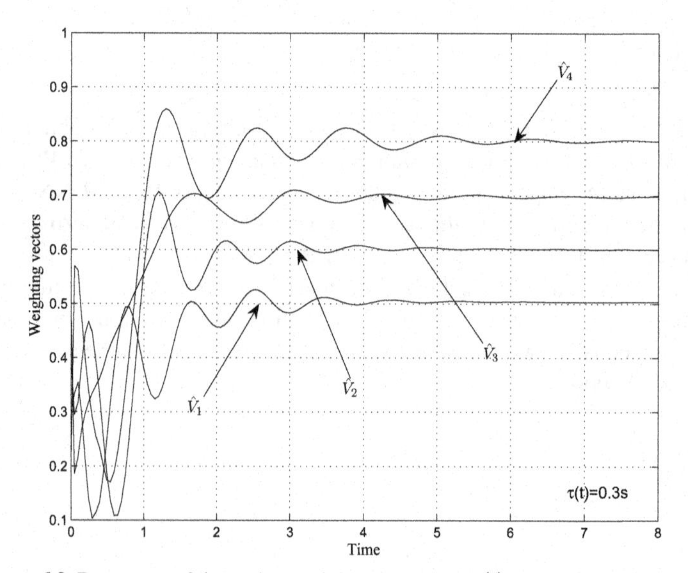


**Figure 6.9** Responses of dynamic weighting vectors for $\tau(t) = 0.3$ s

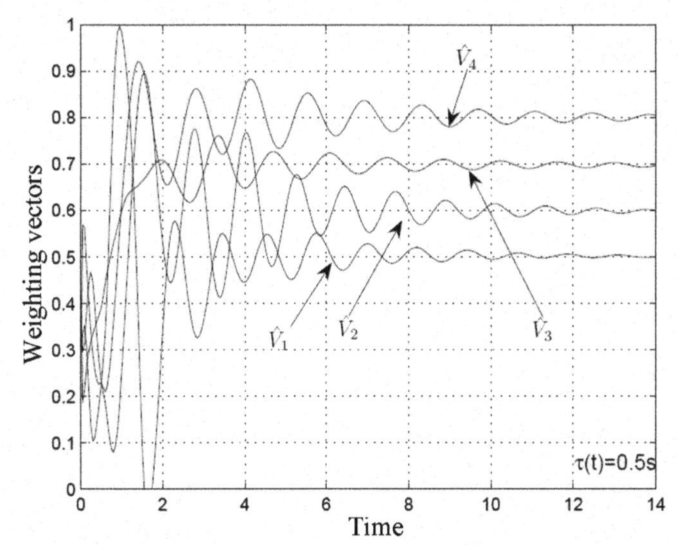


**Figure 6.10** Responses of dynamic weighting vectors for $\tau(t) = 0.5$ s

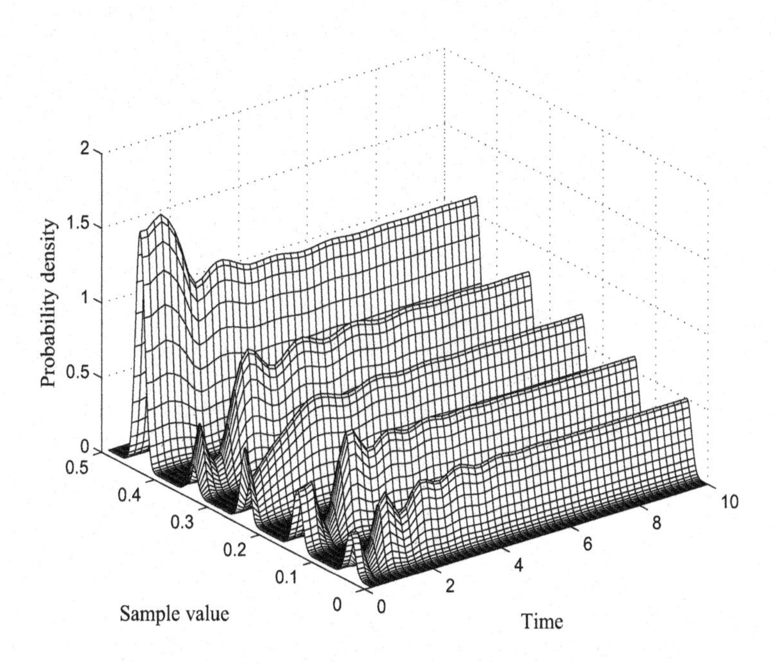


**Figure 6.11** 3D mesh plot of the output PDFs

# Chapter 7
# Constrained Proportional Integral Tracking Control for Takagi-Sugeno Fuzzy Model

## 7.1 Introduction

Similarly to previous PDF tracking problems, after using B-spline approximation theory for the output PDFs, the SDC problem can be transformed into a tracking problem for weighting dynamics. A T-S fuzzy model, as a model identifier, is first utilized to identify the nonlinear dynamics that cannot be modeled exactly, which represents a significant extension to the previous results. A robust tracking problem is studied for a T-S fuzzy weight system which models non-zero equilibriums, time delays, partial state constraints and exogenous disturbances. Meanwhile, a PI controller can be obtained through solving improved LMIs such that stability, tracking performances, robustness and state constraints can be guaranteed simultaneously. Furthermore, the peak-to-peak measure is applied to optimize the tracking performance, which generalizes the corresponding result for linear systems with zero equilibrium.

## 7.2 Problem Statement and Preliminaries

### *7.2.1 Output PDF Model Using B-spline Expansion*

For a dynamic stochastic system, denote $u(t) \in R^m$ as the control input, $z(t) \in [a,b]$ as the stochastic output, then the probability of output $z(t)$ lying inside $[a,y]$ can be described by

$$P(a \leq z(t) < y, u(t)) = \int_a^y \gamma(\eta, u(t)) \mathrm{d}\eta \tag{7.1}$$

where $\gamma(y,u(t))$ is the PDF of the stochastic variable $y(t)$ under control input $u(t)$. For $\sqrt{\gamma(y,u(t))}$, the square root B-spline NNs are given as

$$\sqrt{\gamma(y,u(t))} = \sum_{i=1}^{n} \upsilon_i(u(t))B_i(y) + \varepsilon(y,t) \tag{7.2}$$

where $B_i(y)(i = 1,\cdots,n)$ are pre-specified basis functions and $\upsilon_i(t) := \upsilon_i(u(t))(i = 1,2,\cdots,n)$ are the corresponding weights. Since Equation 7.2 means

$$\gamma(y,u(t)) = (\sum_{i=1}^{n} \upsilon_i(u(t))B_i(y) + \varepsilon(y,t))^2 \geq 0, \forall y \in [a,b]$$

it can be seen that the positiveness of $\gamma(y,u(t))$ can be automatically guaranteed. On the other hand, the PDF should satisfy the condition $\int_a^b \gamma(y,u(t))\mathrm{d}y = 1$, which means only $n-1$ weights are independent. So square root expansions are considered as follows:

$$\gamma(y,u(t)) = (C_0(y)V_0(t) + \upsilon_n(t)B_n(y) + \varepsilon(y,t))^2 \tag{7.3}$$

where

$$C_0(y) = [B_1(y) \;\; B_2(y) \;\; \cdots \;\; B_{n-1}(y)],$$
$$V_0(t) = [\upsilon_1(t) \;\; \upsilon_2(t) \;\; \cdots \;\; \upsilon_{n-1}(t)]^T.$$

In order to fulfill PDF tracking, $\varepsilon(y,t)$ will be assumed to be given by $\varepsilon(y,t) = C_0(y)w_0(t)$, where $w_0(t)$ can be regarded as an unknown perturbation. Hence, Equation 7.3 can be rewritten as

$$\gamma(y,u(t)) = (C_0(y)V(t) + \upsilon_n(t)B_n(y))^2, \quad V(t) = V_0(t) + w_0(t). \tag{7.4}$$

For simplicity, we denote

$$\Lambda_1 = \int_a^b C_0^T(y)C_0(y)\mathrm{d}y, \;\; \Lambda_2 = \int_a^b C_0(y)B_n(y)\mathrm{d}y, \;\; \Lambda_3 = \int_a^b B_n^2(y)\mathrm{d}y. \tag{7.5}$$

To guarantee $\int_a^b \gamma(y,u(t))\mathrm{d}y = 1$, $V^T(t)\Lambda_2^T\Lambda_2V(t) - (V^T(t)\Lambda_1V(t) - 1)\Lambda_3 \geq 0$ should be satisfied, which is equivalent to

$$V^T(t)T_0V(t) \leq 1 \tag{7.6}$$

where $T_0 = \Lambda_1 - \Lambda_3^{-1}\Lambda_2^T\Lambda_2 > 0$. Under condition (7.6), $\upsilon_n$ can be represented by a function of $V(t)$ (see [11] for the details):

$$\upsilon_n(t) = h(V(t)) = \frac{\sqrt{\Lambda_3 - V^T(t)\Lambda_0V(t)} - \Lambda_2V(t)}{\Lambda_3} \tag{7.7}$$

where $\Lambda_0 = \Lambda_1\Lambda_3 - \Lambda_2^T\Lambda_2$. Inequality (7.6) can be considered as a constraint on $V(t)$, which forms one of the key difficulties in the controller design.

Corresponding to Equation 7.3, a desired (known) PDF to be tracked can be described by

$$g(y) = (C_0(y)V_g + h(V_g)B_n(y))^2 \tag{7.8}$$

where $V_g$ is the desired weighting vectors with respect to the same $B_i(y)$. The tracking objective is to find $u(t)$ such that $\gamma(y,u(t))$ can follow $g(y)$. The error between the output PDF and the target PDF is formulated by $\Delta_e = \sqrt{\gamma(y,u(t))} - \sqrt{g(y)}$, i.e.,

$$\Delta_e = C_0 e(t) + [h(V(t)) - h(V_g)]B_n(y) \tag{7.9}$$

where $e(t) = V(t) - V_g$. Due to the continuity of $h(V(t))$, $\Delta_e \to 0$ holds as long as $e(t) \to 0$. The PDF control problem can be transformed into the tracking problem for the above nonlinear weight systems, and the control objective is to find $u(t)$ such that the tracking performance, state constraints and stability are guaranteed simultaneously.

### 7.2.2 PI Controller Design Based on T-S Fuzzy Model

Recently, the T-S fuzzy model has been proved to be a very good representation for a certain class of nonlinear dynamics in control systems and signal processing. So we consider the nonlinear weight dynamics, which could be described by the following T-S fuzzy model with $r$ plant rules:

**Plant Rule** $i$: If $\theta_1$ is $\mu_{i1}$ and $\cdots$ and $\theta_p$ is $\mu_{ip}$, then

$$\dot{V}(t) = A_{0i}V(t) + F_{0i}V_\tau(t) + B_{01i}u(t) + B_{02i}u_\tau(t) + E_{0i}w(t) \tag{7.10}$$

where $V(t) \in R^{n-1}$ is the independent weight vector, $u(t)$ and $w(t)$ represent the control input and the exogenous perturbation, respectively. $w(t)$ is assumed to satisfy $\|w\|_\infty = \sup_{t\geq 0}\|w(t)\| < \infty$. $V_\tau(t) := V(t-\tau(t))$ represent the time delay weight vectors. $u_\tau(t) := u(t-\tau(t))$ is the control input with time delay term. $A_{0i}, F_{0i}, B_{01i}, B_{02i}$ and $E_{0i}$ are known coefficient matrices with compatible dimensions. $\theta_j(x)$ and $\mu_{ij}$ $(i = 1,\cdots,r, j = 1,\cdots,p)$ are, respectively, the premise variables and the fuzzy sets, $r$ is the number of If-Then rules, and $p$ is the number of premise variables. The time-varying delay $\tau(t)$ is assumed to satisfy $0 < \dot{\tau}(t) < \beta < 1$.

By fuzzy blending, the overall fuzzy model is inferred as follows:

$$\begin{aligned}\dot{V}(t) &= \frac{\sum_{i=1}^{r}\omega_i(\theta)(A_{0i}V(t) + F_{0i}V_\tau(t) + B_{01i}u(t) + B_{02i}u_\tau(t) + E_{0i}w(t))}{\sum_{i=1}^{r}\omega_i(\theta)} \\ &= \sum_{i=1}^{r} h_i(\theta)(A_{0i}V(t) + F_{0i}V_\tau(t) + B_{01i}u(t) + B_{02i}u_\tau(t) + E_{0i}w(t))\end{aligned} \tag{7.11}$$

where $\theta = [\theta_1,\cdots,\theta_p]$, $\omega_i : R^p \to [0,1], i = 1,\cdots,r$ is the membership function of the system with respect to plant rule $i$, and $h_i(\theta) = \omega_i(\theta)/\sum_{i=1}^{r}\omega_i(\theta)$. It is obvious that $h_i(\theta) \geq 0$ and $\sum_{i=1}^{r} h_i(\theta) = 1$ can be guaranteed. Based on the above mentioned T-S fuzzy model (7.11), we introduce a new state variable $x(t) := [V^T(t), \int_0^t e^T(\tau)d\tau]^T$. Then the following augmented system with disturbance $w(t)$ and reference input

$V_g(t)$ can be established:

$$\begin{cases} \dot{x}(t) = \sum_{i=1}^{r} h_i(\theta)(A_i x(t) + F_i x_\tau(t) + B_{1i}u(t) + B_{2i}u_\tau(t) + E_i w(t) + HV_g(t)) \\ z(t) = \sum_{i=1}^{r} h_i(\theta)(C_i x(t) + D_i w(t)) \\ x(t) = \phi(t), \quad t \in [-\tau(t), 0] \end{cases} \tag{7.12}$$

where $z(t)$ is the controller output, $\phi(t)$ represents the initial condition of system (7.12),

$$A_i = \begin{bmatrix} A_{0i} & 0 \\ I & 0 \end{bmatrix}, \quad F_i = \begin{bmatrix} F_{0i} & 0 \\ 0 & 0 \end{bmatrix}, \quad B_{1i} = \begin{bmatrix} B_{01i} \\ 0 \end{bmatrix},$$

$$B_{2i} = \begin{bmatrix} B_{02i} \\ 0 \end{bmatrix}, \quad E_i = \begin{bmatrix} E_{0i} \\ 0 \end{bmatrix}, \quad H = \begin{bmatrix} 0 \\ -I \end{bmatrix}.$$

To solve the tracking problem for weight vectors, a direct PI controller is given as

**Plant Rule** $j$: If $\theta_1$ is $\mu_{j1}$ and $\cdots$ and $\theta_p$ is $\mu_{jp}$, then

$$u(t) = \sum_{j=1}^{r} h_j(\theta)(K_{Pj}V(t) + K_{Ij}\int_0^t e(\tau)\mathrm{d}\tau), \quad j = 1, \cdots, r \tag{7.13}$$

where $K_{Pj}$ and $K_{Ij}$ are controller gains to be determined.

With such an augmented system (7.12), the tracking problem can be further reduced to a stabilization control framework because the PI controller can be formulated as

$$u(t) = \sum_{j=1}^{r} h_j(\theta)[K_j x(t)], \quad K_j = [K_{Pj} \ \ K_{Ij}]. \tag{7.14}$$

## 7.3 Main Results

### *7.3.1 Stability Analysis with $L_1$ Measure Index*

Since $V_g$ can be seen as a known vector, we denote $y_d := \|V_g\|^2$. The following result provides a criterion for the $L_1$ performance problem of the unforced system (7.12), which also generalizes the corresponding result for linear systems with zero equilibrium in the absence of state constraints.

**Theorem 7.1** For the known parameters $\mu_i (i = 1, 2, 3)$ , $\alpha > 0$ and $\gamma > 0$, suppose that there exist $S, P > 0$ and for $i = 1, \cdots, r$, $j = 1, \cdots, r$ such that the following LMIs

$$\begin{bmatrix} \mathrm{sym}(A_i^T P) + \mu_1^2 P + S & PF_i & PE_i & PH \\ F_i^T P & -(1-\beta)S & 0 & 0 \\ E_i^T P & 0 & -\mu_2^2 I & 0 \\ H^T P & 0 & 0 & -\mu_3^2 I \end{bmatrix} < 0, \tag{7.15}$$

$$\begin{bmatrix} \mu_1^2 P & 0 & \frac{1}{2}(C_i^T + C_j^T) \\ 0 & (\gamma - \mu_2^2 - \mu_3^2 y_d)I & \frac{1}{2}(D_i^T + D_j^T) \\ \frac{1}{2}(C_i + C_j) & \frac{1}{2}(D_i + D_j) & \gamma I \end{bmatrix} > 0, \tag{7.16}$$

$$\begin{bmatrix} \alpha I & P \\ P & P \end{bmatrix} > 0, \quad \begin{bmatrix} P & 0 & \frac{1}{2}(C_i^T + C_j^T) \\ 0 & (\gamma - \alpha x_m^T x_m)I & \frac{1}{2}(D_i^T + D_j^T) \\ \frac{1}{2}(C_i + C_j) & \frac{1}{2}(D_i + D_j) & \gamma I \end{bmatrix} > 0 \tag{7.17}$$

are solvable, then the unforced system (7.12) is stable, and $\sup_{0 \le \|w\| \le \infty} \frac{\|z(t)\|_\infty}{\|w(t)\|_\infty} < \gamma$ holds.

*Proof.* Define a Lyapunov-Krasovskii function

$$S_1(x(t),t) = x^T(t)Px(t) + \int_{t-\tau(t)}^{t} x^T(\beta)Sx(\beta)\mathrm{d}\beta. \tag{7.18}$$

Obviously, $S_1(x(t),t) \ge 0$. Furthermore, it can be seen that

$$\begin{aligned}\frac{\mathrm{d}S_1(x(t),t)}{\mathrm{d}t} &= 2x^T(t)P\dot{x}(t) + x^T(t)Sx(t) - (1-\dot{\tau}(t))x_\tau^T(t)Sx_\tau(t) \\ &= \sum_{i=1}^{r} h_i(\theta)x^T(t)(PA_i + A_i^T P + S)x(t) - (1-\dot{\tau}(t))x_\tau^T(t)Sx_\tau(t) \\ &\quad +2x^T(t)PHV_g + 2\sum_{i=1}^{r} h_i x^T(t)PE_i w(t) + 2\sum_{i=1}^{r} h_i x^T(t)PF_i x_\tau(t) \\ &\le \sum_{i=1}^{r} h_i(\theta)\zeta^T(t)\Phi_i\zeta(t) + \|\mu_2 w(t)\|^2 + \mu_3^2 y_d \end{aligned} \tag{7.19}$$

where

$$\zeta(t) = [x^T(t),\ x_\tau^T(t)]^T, \quad \Phi_i = \begin{bmatrix} \Upsilon_i & PF_i \\ F_i^T P & -(1-\beta)S \end{bmatrix},$$

$$\Upsilon_i = A_i^T P + PA_i + S + \frac{1}{\mu_2^2} PE_i E_i^T P + \frac{1}{\mu_3^2} PHH^T P. \tag{7.20}$$

Based on the Schur complement formula, LMI (7.15) implies that for any $w(t)$ satisfying $\|w(t)\|_\infty \le 1$, $\Phi_i < \begin{bmatrix} -\mu_1^2 P & 0 \\ 0 & 0 \end{bmatrix}$ holds. With inequality (7.19), it can be seen that

$$\frac{\mathrm{d}S_1(x(t),t)}{\mathrm{d}t} \le -\mu_1^2 x^T(t)Px(t) + \mu_2^2 + \mu_3^2 y_d \tag{7.21}$$

where $\mu_3^2 y_d$ can be considered a known parameter. Thus, $\frac{\mathrm{d}S_1(x(t),t)}{\mathrm{d}t} < 0$, if $x^T(t)Px(t) > \mu_1^{-2}(\mu_2^2 + \mu_3^2 y_d)$ holds. So for any $x(t)$, it can be verified that

$$x^T(t)Px(t) \le \max\{x_m^T P x_m, \mu_1^{-2}(\mu_2^2 + \mu_3^2 y_d)\},$$

$$\|x_m\| = \sup_{-\tau(t)\le t\le 0}\|x(t)\|, \tag{7.22}$$

which also implies that the unforced system (7.12) is stable.

From inequality (7.22) we can get that $x^T(t)Px(t) \le x_m^T P x_m$ or $x^T(t)Px(t) \le \mu_1^{-2}(\mu_2^2+\mu_3^2 y_d)$. By defining $\eta(t) = [x^T(t), w^T(t)]^T$, $H_i = [C_i, D_i]$ and $H_j = [C_j, D_j]$, we have

$$\|z(t)\|^2 = \Sigma_{i=1}^r \Sigma_{j=1}^r h_i(\theta)h_j(\theta)(C_i x(t) + D_i w(t))^T (C_j x(t) + D_j w(t))$$

$$\le \frac{1}{4}\Sigma_{i=1}^r \Sigma_{j=1}^r h_i(\theta)h_j(\theta)\eta^T(t)(H_i+H_j)^T(H_i+H_j)\eta(t).$$

From LMI (7.16), it can be seen that

$$\begin{bmatrix} \mu_1^2 P & 0 \\ 0 & (\gamma-\mu_2^2-\mu_3^2 y_d)I \end{bmatrix} - \frac{1}{4\gamma}\begin{bmatrix} C_i^T + C_j^T \\ D_i^T + D_j^T \end{bmatrix}\begin{bmatrix} C_i + C_j & D_i + D_j \end{bmatrix} > 0,$$

which guarantees under $x^T(t)Px(t) \le \mu_1^{-2}(\mu_2^2+\mu_3^2 y_d)$ and $\|w(t)\|_\infty \le 1$ that we obtain

$$\frac{1}{\gamma}\|z(t)\|^2 < (\mu_2^2+\mu_3^2 y_d) + (\gamma-\mu_2^2-\mu_3^2 y_d)w^T(t)w(t) = \gamma. \tag{7.23}$$

On the other hand, from LMI (7.17), it can also be shown that

$$\begin{bmatrix} P & 0 \\ 0 & (\gamma-\alpha x_m^T x_m)I \end{bmatrix} - \frac{1}{4\gamma}\begin{bmatrix} C_i^T + C_j^T \\ D_i^T + D_j^T \end{bmatrix}\begin{bmatrix} C_i + C_j & D_i + D_j \end{bmatrix} > 0.$$

Similarly to the above proof, under $x^T(t)Px(t) \le x_m^T P x_m$ and $\|w(t)\|_\infty \le 1$, we obtain

$$\frac{1}{\gamma}\|z(t)\|^2 < \alpha x_m^T x_m + (\gamma-\alpha x_m^T x_m)w^T(t)w(t) = \gamma. \tag{7.24}$$

Hence, the $L_1$ norm of the unforced system is less than $\gamma$. □

### 7.3.2 Peak-to-peak Tracking Performance

Considering the state feedback controller with PI control structure, and substituting $u(t) = \Sigma_{j=1}^r h_j(\theta)[K_j x(t)]$ into Equation 7.12, the corresponding nonlinear closed-loop system can be described by

$$\begin{cases} \dot{x}(t) = \Sigma_{i=1}^r h_i(\theta)\Sigma_{j=1}^r h_j(\theta)[(A_i+B_{1i}K_j)x(t) + (F_i+B_{2i}K_j)x_\tau(t) \\ \qquad\qquad +E_i w(t) + HV_g(t)] \\ z(t) = \Sigma_{i=1}^r h_i(\theta)[C_i x(t) + D_i w(t)]. \end{cases} \tag{7.25}$$

The following results provide a solution for the nonlinear tracking control problem with disturbance attenuation performance.

**Theorem 7.2** For the known parameters $\mu_i (i = 1,2,3)$ , $\alpha > 0$ and $\gamma > 0$, suppose that there exist $S > 0$, $Q = P^{-1}$ and for $i, j = 1, \cdots, r$ such that the following LMIs

$$\begin{bmatrix} \mathrm{sym}(A_i Q) + \mathrm{sym}(B_{1i} R_j) + \mu_1^2 Q + QSQ & F_i Q + B_{2i} R_j & E_i & H \\ QF_i^T + R_j^T B_{2i}^T & -(1-\beta) QSQ & 0 & 0 \\ E_i^T & 0 & -\mu_2^2 I & 0 \\ H^T & 0 & 0 & -\mu_3^2 I \end{bmatrix} < 0, \tag{7.26}$$

$$\begin{bmatrix} \mu_1^2 Q & 0 & \frac{1}{2}(QC_i^T + QC_j^T) \\ 0 & (\gamma - \mu_2^2 - \mu_3^2 y_d) I & \frac{1}{2}(D_i^T + D_j^T) \\ \frac{1}{2}(C_i Q + C_j Q) & \frac{1}{2}(D_i + D_j) & \gamma I \end{bmatrix} > 0, \tag{7.27}$$

$$\begin{bmatrix} \alpha I & I \\ I & Q \end{bmatrix} > 0, \quad \begin{bmatrix} Q & 0 & \frac{1}{2}(QC_i^T + QC_j^T) \\ 0 & (\gamma - \alpha x_m^T x_m) I & \frac{1}{2}(D_i^T + D_j^T) \\ \frac{1}{2}(C_i Q + C_j Q) & \frac{1}{2}(D_i + D_j) & \gamma I \end{bmatrix} > 0 \tag{7.28}$$

are solvable, then the closed-loop system (7.25) is stable and satisfies both $\lim_{t \to \infty} V(t) = V_g$ and $\sup_{0 \le \|w\| \le \infty} \frac{\|z(t)\|_\infty}{\|w(t)\|_\infty} < \gamma$. The PI control gain $K_j$ can be solved via $R_j = K_j Q$.

*Proof.* Based on Theorem 7.1 and Lyapunov-Krasovskii function (7.18), we obtain

$$\begin{aligned} \frac{dS_1(x(t),t)}{dt} &= \sum_{i=1}^{r} h_i \sum_{j=1}^{r} h_j x^T(t)(PA_i + A_i^T P + PB_{1i}K_j + K_j^T B_{1i}^T P + S)x(t) \\ &\quad + 2\sum_{i=1}^{r} h_i \sum_{j=1}^{r} h_j x^T(t) PB_{2i} K_j x_\tau(t) - (1 - \dot{\tau}(t)) x_\tau^T(t) S x_\tau(t) \\ &\quad + 2\sum_{i=1}^{r} h_i x^T(t) PF_i x_\tau(t) + 2x^T(t) PHV_g + 2\sum_{i=1}^{r} h_i x^T(t) PE_i w(t) \\ &\le \sum_{i=1}^{r} h_i \sum_{j=1}^{r} h_j \zeta^T(t) \Omega_{ij} \zeta(t) + \|\mu_2 w(t)\|^2 + \mu_3^2 y_d \end{aligned} \tag{7.29}$$

where

$$\zeta = [x^T(t),\ x_\tau^T(t)]^T, \quad \Omega_{ij} = \begin{bmatrix} \Xi_{ij} & PF_i + PB_{2i}K_j \\ F_i^T P + K_j^T B_{2i}^T P & -(1-\beta)S \end{bmatrix},$$

$$\Xi_{ij} = \mathrm{sym}(A_i^T P) + \mathrm{sym}(K_j^T B_{1i}^T P) + S + \frac{1}{\mu_2^2} PE_i E_i^T P + \frac{1}{\mu_3^2} PHH^T P. \tag{7.30}$$

Based on the Schur complement formula, by pre-multiplying $\mathrm{diag}\{Q^{-1}, I, I, I\}$ and post-multiplying $\mathrm{diag}\{I, Q^{-1}, I, I\}$ on both side of LMI (7.26), we can get $\Omega_{ij} \le$

$\text{diag}\{-\mu_1^2, P, 0\}$. So for any $w(t)$ satisfying $\|w(t)\|_\infty \leq 1$, it can be seen that

$$\frac{\mathrm{d}S_1(x(t),t)}{\mathrm{d}t} \leq -\mu_1^2 x^T(t)Px(t) + \mu_2^2 + \mu_3^2 y_d. \tag{7.31}$$

Similarly to the proof of Theorem 7.1, it can be seen that inequality (7.22) still holds for the closed-loop system, which implies that system (7.25) is stable in the presence of $w(t)$ and $V_g$. Meanwhile, the closed-loop system satisfies the peak-to-peak performance through LMIs (7.27) and (7.28).

For a couple of $w(t)$ and $V_g$, we suppose that $\vartheta_1(t)$ and $\vartheta_2(t)$ are two trajectories of the closed-loop system corresponding to a fixed initial condition. Then the variable $\sigma(t) := \vartheta_1(t) - \vartheta_2(t)$ satisfies the following dynamic equation:

$$\dot{\sigma}(t) = \sum_{i=1}^{r} h_i(\theta) \sum_{j=1}^{r} h_j(\theta)[(A_i + B_{1i}K_j)\sigma(t) + (F_i + B_{2i}K_j)\sigma_\tau(t)]. \tag{7.32}$$

Similarly to Equation 7.18, a Lyapunov function can be constructed as

$$S_2(\sigma(t),t) = \sigma^T(t)P\sigma(t) + \int_{t-\tau(t)}^{t} \sigma_\tau^T(\beta)S\sigma_\tau(\beta)\mathrm{d}\beta. \tag{7.33}$$

From LMI (7.26), it can be seen that

$$\frac{\mathrm{d}S_2(\sigma(t),t)}{\mathrm{d}t} \leq -\mu_1^2 \sigma^T(t)P\sigma(t) \leq -\mu_1^2 \lambda_{min}(P)\|\sigma(t)\|^2 < 0 \tag{7.34}$$

where $\lambda_{min}(P)$ is the minimal eigenvalue of $P$. It can be verified that $\sigma = 0$ is the unique asymptotically stable equilibrium point of the system (7.32). This means that the closed-loop system (7.25) also has a unique asymptotically stable equilibrium point. It can be concluded that $\lim_{t\to\infty} \frac{\mathrm{d}}{\mathrm{d}t}(\int_0^t e(\tau)\mathrm{d}\tau) = 0$, which shows that $\lim_{t\to\infty} V(t) = V_g$. □

### 7.3.3 Constrained Peak-to-peak Tracking Control

It has been shown that, due to the properties of the PDF, the weight vectors have to satisfy $V^T(t)T_0V(t) \leq 1$, which can be reduced to $x^T(t)Tx(t) \leq 1$, where $T := \text{diag}\{T_0, 0\}$. Theorem 7.2 shows that the equilibrium of system (7.25) is not the origin. The disturbances make the constrained tracking problem more complicated, compared with the previous results [61, 63].

**Theorem 7.3** For the known parameters $\mu_i (i = 1,2,3)$, $\alpha > 0$ and $\gamma > 0$, suppose that there exist $S > 0$, $Q = P^{-1}$ such that LMIs (7.26-7.28) and

$$(\mu_2^2 + \mu_3^2 y_d)T \leq \mu_1^2 P, \tag{7.35}$$

$$T \le P, \quad \begin{bmatrix} 1 & x_m^T P \\ P x_m & P \end{bmatrix} \ge 0 \tag{7.36}$$

are solvable, then closed-loop system (7.25) is stable, satisfies $x^T(t)Tx(t) \le 1$, $\lim_{t\to\infty} V(t) = V_g$ and $\sup_{0\le\|w\|\le\infty} \frac{\|z(t)\|_\infty}{\|w(t)\|_\infty} < \gamma$. The PI control gain $K_j$ can be solved via $R_j = K_j Q$.

*Proof.* Based on Theorem 7.2, it can be shown that the stability, tracking performance and disturbance attenuation performance for the closed-loop are guaranteed. Remaining is to show that under conditions (7.35) and (7.36), the closed-loop system satisfies the desired constraints with respect to PDF control.

Similarly to the above proof, inequality (7.22) still holds. This means that $x^T(t)Px(t) \le x_m^T P x_m$ or $x^T(t)Px(t) \le \mu_1^{-2}(\mu_2^2 + \mu_3^2 y_d)$. Combining with condition (7.36) it can be shown that

$$x^T(t)Tx(t) \le x^T(t)Px(t) \le x_m^T P x_m \le 1. \tag{7.37}$$

On the other hand, from condition (7.35) it can be seen that

$$x^T(t)Tx(t) \le \mu_1^2(\mu_2^2 + \mu_3^2 y_d)^{-1} x^T(t)Px(t) \le 1. \tag{7.38}$$

□

## 7.4 An Illustrative Example

Suppose that the output PDFs can be approximated using the square root B-spline models described by Equation 7.3 with $n = 3$, $y \in [0, 1.5]$, $i = 1,2,3$

$$B_i(y) = \begin{cases} |\sin 2\pi y| & y \in [0.5(i-1); 0.5i] \\ 0 & y \in [0.5(j-1); 0.5j] \quad i \ne j. \end{cases}$$

From the notation in Equation 7.5, it can be seen that $\Lambda_1 = \mathrm{diag}\{0.25, 0.25\}, \Lambda_2 = [0, 0]$, $\Lambda_3 = 0.25$. The desired PDF $g(y)$ is assumed to be described by Equation 7.8 with $V_g = [0.6, 1]^T$.

Consider a T-S fuzzy model with time delay term and exogenous perturbation; the model parameters are given as follows:

$$A_{01} = \begin{bmatrix} -3 & 3 \\ -2 & 3 \end{bmatrix},\ F_{01} = \begin{bmatrix} 0.3 & 0 \\ 0.3 & 0 \end{bmatrix},\ B_{011} = \begin{bmatrix} -2 & 0 \\ 0 & 2 \end{bmatrix},\ B_{012} = \begin{bmatrix} 0 & 1 \\ 1 & 0 \end{bmatrix},\ E_{01} = \begin{bmatrix} 1 & 0 \\ 0 & 1 \end{bmatrix},$$

$$A_{02} = \begin{bmatrix} -2 & 2 \\ -1 & 2 \end{bmatrix}, F_{02} = \begin{bmatrix} -2 & 0 \\ -1 & -1 \end{bmatrix}, B_{021} = \begin{bmatrix} 1 & 0 \\ 0 & -1 \end{bmatrix}, B_{022} = \begin{bmatrix} 0.3 & 0 \\ 0 & 0.3 \end{bmatrix}, E_{02} = \begin{bmatrix} 1 & 0 \\ 0 & 1 \end{bmatrix}.$$

In this example, we choose Gaussian functions as our member functions, which are

$$M_1 = \frac{e^{\left(\frac{-(x_2+1)^2}{\sigma^2}\right)}}{e^{\left(\frac{-(x_2+1)^2}{\sigma^2}\right)} + e^{\left(\frac{-(x_2-1)^2}{\sigma^2}\right)}}, \quad M_2 = \frac{e^{\left(\frac{-(x_2-1)^2}{\sigma^2}\right)}}{e^{\left(\frac{-(x_2+1)^2}{\sigma^2}\right)} + e^{\left(\frac{-(x_2-1)^2}{\sigma^2}\right)}}$$

where $\sigma = 0.8$ and $w(t)$ is assumed as white noise, meanwhile, we define $\alpha = \mu_1^2 = 1$, $\mu_2^2 = \mu_2^2 = 3$, $\lambda = 8$ and the time delay $\tau(t) = 0.2$. Through solving the LMIs (7.26)-(7.28) and (7.35)-(7.36), the PI control gains (7.13) can be computed as follows:

$$K_{P1} = \begin{bmatrix} 25.2073 & -0.5814 \\ -3.0595 & -32.0404 \end{bmatrix}, \quad K_{I1} = \begin{bmatrix} 30.5210 & -4.7896 \\ -5.0719 & -33.1997 \end{bmatrix},$$

$$K_{P2} = \begin{bmatrix} 23.9925 & -26.0358 \\ -25.6907 & -28.6723 \end{bmatrix}, \quad K_{I2} = \begin{bmatrix} 27.6868 & -31.4657 \\ -31.7217 & -25.9541 \end{bmatrix}.$$

The responses of dynamical weighting vectors are shown in Figure 7.1. Under the proposed robust control strategy, the 3D mesh plot of practical PDF is shown in Figure 7.2.

## 7.5 Conclusions

This chapter considers the robust tracking problem for the output PDF of non-Gaussian processes using a generalized PI controller. The main results here have four features:

1. The T-S fuzzy model, as a system identifier, is first applied into stochastic distribution problem.
2. Exogenous disturbance, state constraint and non-zero equilibrium are all considered in the T-S fuzzy model tracking control problem.
3. Using LMI methods, multiple control objectives including stabilization, tracking performance, robustness and state constraint can be guaranteed simultaneously.
4. To enhance robustness, the peak-to-peak measure is applied to optimize tracking performance.

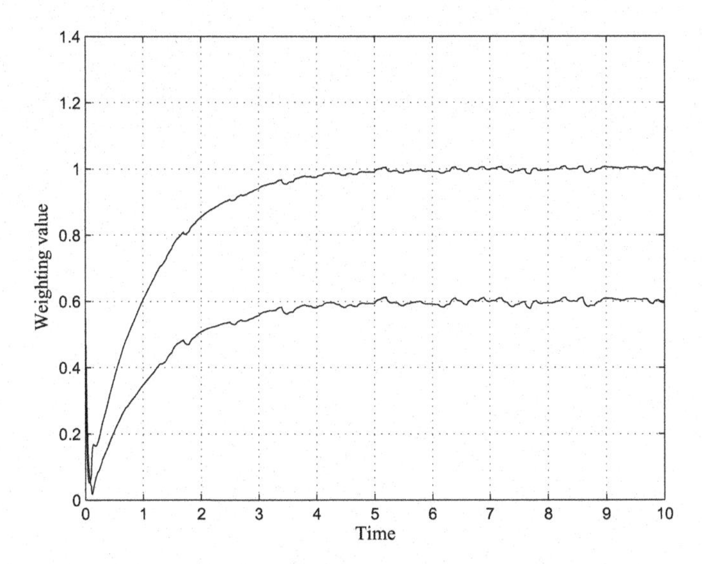


**Figure 7.1** Responses of weighting vectors

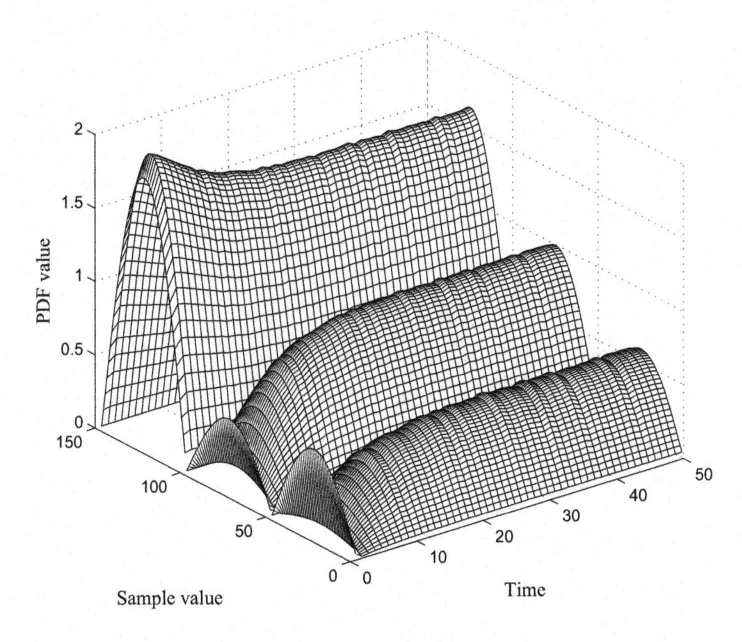


**Figure 7.2** 3D mesh plot of the PDFs

# Part III
# Statistical Tracking Control – Driven by Output Statistical Information Set

Gaussian systems have received much attention in the past decades, where mean and variance are the objectives for control and filtering (see [3, 7, 18, 54, 180]). In the classical stochastic tracking control theory, the objective is to ensure the mean or variance of the difference between the output and a given reference signal is either convergent or minimized [53, 137]. non-Gaussian variables exist in many complex stochastic systems, which may even have asymmetric and multiple peak stochastic distributions (see [66, 157] for details). For non-Gaussian systems, mean and variance are insufficient to characterize the stochastic properties.

Along with the rapid development of sensor and computer technology, the measured information or system knowledge in some processes comprises huge amounts of data or images, from which statistical information (including the moments and the entropy) or PDFs can be characterized using different data processing or imagine processing techniques. Similarly to previous discussions, this kind of problem has been called PDF tracking or shape control, or the SDC problem (see [63, 64, 163, 163]). However, it is difficult to determine the output PDFs using analytical methods due to the high nonlinearity of both the nonlinear stochastic system and the PDFs. It is shown that for Itô equations, Fokker-Planck-Kolmogorov equations have to be formulated to describe the state PDFs and numeral approaches have to been used (see [25, 34, 39, 138]). In order to obtain some feasible controller design algorithms for the SDC problem, B-spline NN have been introduced to model the dynamic relationships between the control input and the output PDFs so that the problem can be reduced to a tracking problem in a finite-dimension scenario for the weight systems. More importantly, some fundamental theoretical issues can be addressed formally, such as controllability, control accuracy and convergent rate. As shown in [63, 64], it is a challenging issue to reduce the problem complexity and to supply feasible control laws with satisfactory closed-loop performance.

One possible question for the SDC problem is how to assess the controllability when a finite dimensional input is used to control the constrained PDF. In this part, we will present some results on the data-driven STC problem for stochastic distribution systems, and name it statistical information tracking control. Differing from either the conventional stochastic tracking or the recent PDF tracking problems, the information for feedback is the set of the statistical information of the output, and the goal of control is to ensure the statistical information of the system output follows that of a target set. Specifically, the mean and entropy have been used as the tracked statistical information. In comparison with the results for PDF tracking, the main results in this part have two features. First, since the mean and the variance are two commonly used control objectives for Gaussian systems, our control objective is a reasonable generalization for non-Gaussian systems. Indeed, the objective is the tracking of the target PDF, which can be achieved by statistical information tracking. Second, the obtained statistical tracking will eliminate the constraints (see [64, 66, 157]) widely seen in B-spline NN approximations of the output PDFs.

# Chapter 8
# Multiple-objective Statistical Tracking Control Based on Linear Matrix Inequalities

## 8.1 Introduction

A new type of control framework for non-Gaussian stochastic systems called statistical tracking control (STC) is established in this chapter. Different from both the traditional stochastic optimization objective for Gaussian systems and the PDF tracking objective for non-Gaussian systems, the tracking objective is the set of statistical information (including the moments and the entropy) of a given target PDF, rather than a deterministic signal. The control aims at making the statistical information of the system output PDFs follow those of a target PDF. After using B-spline NN approximation theory for the integrated performance functions in the context of the entropy and mean, it is shown that the above STC problem can be transformed into a tracking problem for the weight vectors. We will consider the generalized mixed $H_2$ and $H_\infty$ controller design for the tracking problem of the time delayed weighting models. The control objective is to find the gains of the mixed $H_2/H_\infty$ controller such that the closed-loop system is asymptotically stable, the weight dynamics can follow the desired set of weights, and the disturbance or un-modeling errors can be attenuated. LMI-based design approaches are provided, which has also independent significance for the tracking problem in the $\mathrm{H}_2/H_\infty$ optimization setting for uncertain time delay systems. Finally, for a B-spline NN model from the paper making process, simulations are provided to show the effectiveness of the proposed results.

## 8.2 Problem Formulation and Preliminaries

For a dynamic stochastic system, denote $u(t) \in R^m$ as the control input, $z(t) \in [a,b]$ as the stochastic output, whose PDF is denoted by $\gamma(y,u(t))$, where $y$ is the variable in the PDF definition interval $[\alpha,\beta]$. It is noted that $\gamma(y,u(t))$ is a dynamic function of $y$ along with the time variable $t$. In previous work, the PDF tracking problem has been studied with some effective design algorithms, where the B-spline expansions have been used to approximate $\gamma(y,u(t))$ or $\sqrt{\gamma(y,u(t))}$ and the control objective is to find $u(t)$ such that $\gamma(y,u(t))$ converges to the target PDF $g(y)$. For instance, in [64], the square root B-spline model is used and the control objective is to minimize $e(y,t) = \sqrt{g(y)} - \sqrt{\gamma(y,u(t))} \to 0$.

The idea results from a simple observation. It is well known that mean and variance can characterize the stochastic property of a Gaussian variable. Generally, the moments from the lower order to higher order can decide the shape of a non-Gaussian PDF. In addition, entropy has been widely used in information, thermodynamics, communication and control theories as a measure for the average information contained in a given PDF of a stochastic variable [155, 187]. Thus, the PDF tracking can be achieved via the tracking of the above proper statistical information. Indeed, in many industrial processes, with the development of sensing and computer technology, the observed information can be described by the statistical information (including moments and entropy) and PDF with respect to the output.

To illustrate the design algorithm, in this chapter we consider a special STC problem. The considered performance index is $\int_{\alpha}^{\beta} \delta(\gamma,u(t))dy$, where

$$\delta(\gamma,u(t)) = Q_1\gamma(y,u(t))\ln(\gamma(y,u(t))) + Q_2 y\gamma(y,u(t)) \tag{8.1}$$

where $Q_1$ and $Q_2$ are two weight parameters. In Equation 8.1 the integral of the first term is the entropy and that of the second one is the mean of the output PDF.

In the following we construct B-spline NN expansions for $\delta(\gamma,u(t))$ as follows

$$\delta(\gamma,u(t)) = C(y)V_0(t) + \varepsilon(y,t) \tag{8.2}$$

where

$$C(y) = [B_1(y)B_2(y)\ldots B_n(y)],$$
$$V_0(t) = [v_1(t)v_2(t)\ldots v_n(t)]^T.$$

$B_i(y)$ $(i = 1,2,\cdots,n)$ are pre-specified basis functions and $v_i(u(t)) := v_i(t), (i = 1,2,\cdots,n)$ are the corresponding weights. $\varepsilon(y,t)$ represents the modeling error. Assuming that $\varepsilon(y,t)$ can be replaced by $\varepsilon(y,t) = C(y)\varepsilon_0(t)$, where $\varepsilon_0(t)$ is also regarded as an unknown perturbation. Hence, Equation 8.2 can be rewritten as

$$\delta(\gamma,u(t)) = C(y)V_0(t) + C(y)\varepsilon_0(t) = C(y)V(t) \tag{8.3}$$

where $V(t) = V_0(t) + \varepsilon_0(t)$. Based on Equation 8.1 and 8.3, for the target PDF (in many cases it can be (but not required) a Gaussian one), we can find the corresponding weights, which are denoted $V_g(t)$. That is, $\delta(g,u(t)) = C(y)V_g(t)$. The tracking objective is to find $u(t)$ such that $\delta(\gamma,u(t)) - \delta(g,u(t)) = C(y)x(t)$ converges to 0, where $x(t) := V(t) - V_g(t)$.

*Remark 8.1* Following B-spline NN approximation to the measured performance function $\delta(\gamma,u(t))$, the problem is transferred into the tracking of weight vectors $V(u(t))$. Obviously the weight vectors are dynamically related to the control input

$u(t)$. Furthermore, the weight errors $x(t)$ are also dynamically related to the control input. Therefore, the next task is to identify the dynamic relationships between the control input and the weight errors $x(t)$.

*Remark 8.2* Compared with tracking control of Gaussian systems, STC can better consider non-Gaussian variables, for instance the entropy and the mean in Equation 5.1. Compared with tracking control of the output PDFs, the constraints (see [64, 66]) resulting from B-spline expansions for the output PDFs can be avoided through STC.

## 8.3 Formulation for the $H_2/H_\infty$ Tracking Problem

Once B-spline expansions have been made for the integrated function (8.1), the next step is to establish the dynamic model between the control input and the weight errors $x(t)$. This procedure has been widely used in PDF control and entropy optimization problems (see [7, 11, 12]). To simplify the design algorithm, originally only linear models are considered, where the shape of the output PDFs cannot be changed [7]. In this chapter, we adopt the following time delay weight error model with nonlinearity and the exogenous disturbances

$$\begin{aligned}\dot{x}(t) = Ax(t) + A_d x(t-d) + Ff(x(t)) + B_{11}w_0(t) + B_{12}V_g \\ + B_2 u(t) + B_{2d}u(t-d) \qquad (8.4)\end{aligned}$$

where $u(t)$ and $w_0(t)$ represent the control input and the exogenous vector, respectively. $d(t)$ is the time delay term satisfying $0 < \dot{d}(t) \le \beta < 1$. The initial condition is assumed to be $x(t) = \phi(t), t \in [-d(t), 0]$. $x(t-d)$ and $u(t-d)$, which represents the delayed state and the delayed control input, respectively. It is noted that here uncertain vector $w_0(t)$ includes two parts. One part can be considered as the unmodeled dynamics of the weighting model (8.4), and another resulted from $\varepsilon(y,t)$. In this chapter, $w_0(t)$ is supposed to satisfy $\|w_0(t)\|_2 < \infty$. $A$, $F$, $A_d$, $B_2$, $B_{2d}$ and $B_{1j}$ $(j = 1,2)$ are known coefficient matrices. $f(x(t))$ is a nonlinear function and satisfies the following Lipschitz property:

$$\|f(x_1(t)) - f(x_2(t))\| \le \|U(x_1(t) - x_2(t))\| \qquad (8.5)$$

for any $x_1(t)$ and $x_2(t)$, where $U$ is a known matrix. It is noted that $f(x(t))$ can also be regarded as a kind of unknown modeling uncertainty. Such a model can also be established using NN identification.

In the context of tracking in finite time, we use the truncated $H_2$ norm $\|x(t)\|_{2T} = \int_{-d}^{T} x^T(\tau)x(\tau)d\tau$ instead of the conventional $H_2$ norm $\|x(t)\|_2$ (see [10]). $H_2$ performance is used to confine the transient response of the system and the high gain of the control input, which can be defined as follows $J_2 := \|z_0(t)\|_{2T}^2$, where the reference output can be selected as $z_0(t) = C_0x(t) + D_0u(t)$, and $C_0$ and $D_0$ are known weight matrices (see [56, 59, 70, 202]) and references therein).

For the weight tracking control problem, we consider the non-zero initial condition and generalization of the conventional $H_\infty$ control. In order to guarantee both tracking performance and disturbance attenuation performance $\|x(t)\|_{2T}^2 \leq \rho^2 \|w_0(t)\|_{2T}^2$, we select the second reference output as

$$z_\infty(t) = \begin{bmatrix} x(t) \\ \rho V_g \end{bmatrix} = \begin{bmatrix} I \\ 0 \end{bmatrix} x(t) + \begin{bmatrix} 0 & 0 \\ 0 & \rho I \end{bmatrix} \begin{bmatrix} w_0(t) \\ V_g \end{bmatrix} \tag{8.6}$$

where $w(t) = \begin{bmatrix} w_0(t) \\ V_g \end{bmatrix}$. It can be shown that $\|z_\infty(t)\|_{2T}^2 \leq \rho^2 \|w(t)\|_{2T}^2$ implies that $\|x(t)\|_{2T}^2 \leq \gamma \|w_0(t)\|_{2T}^2$.

Thus, we obtain the following augmented model for the $H_2/H_\infty$ control:

$$\Sigma_f : \begin{cases} \dot{x}(t) = Ax(t) + A_d x(t-d) + Ff(x(t)) + B_1 w(t) \\ \qquad\qquad + B_2 u(t) + B_{2d} u(t-d) \\ z_0(t) = C_0 x(t) + D_0 u(t) \\ z_\infty(t) = C_1 x(t) + D_1 w(t) \end{cases} \tag{8.7}$$

where $C_1 = \begin{bmatrix} I \\ 0 \end{bmatrix}$, $D_1 = \begin{bmatrix} 0 & 0 \\ 0 & \sqrt{\rho} I \end{bmatrix}$, and $B_1 = [B_{11} \;\; B_{12}]$.

As stated above, in most existing standard $H_\infty$ control and filtering results for delay systems, $\phi(t) = 0$ is assumed for $t \in [-d, 0]$. In order to apply the $H_\infty$ performance measure for non-zero initial conditions, the generalized $H_\infty$ performance measure for $\Sigma_f$ is defined as

$$J_\infty(\gamma, P_0, S_0) := \left( \|z\|_{2T}^2 - \gamma^2 \|w\|_{2T}^2 \right) - \delta(P_0, S_0) \tag{8.8}$$

where

$$\delta(P_0, S_0) := \phi^T(0) P_0 \phi(0) + \int_{-d}^{0} \phi^T(\tau) S_0 \phi(\tau) d\tau.$$

$\gamma > 0$ is a scalar, and $P_0 > 0$, and $S_0 > 0$ are weight matrices.

It is noted that $J_\infty(\gamma, P_0, S_0) < 0$ leads to

$$\|z\|_{2T}^2 \leq \gamma^2 \|w\|_{2T}^2 + \phi^T(0) P_0 \phi(0) + \int_{-d}^{0} \phi^T(\tau) S_0 \phi(\tau) d\tau. \tag{8.9}$$

Thus, if $\phi(t) = 0$ holds for all $t \in [-d, 0]$, the above inequality reduces to $\|z\|_{2T}^2 \leq \gamma^2 \|w\|_{2T}^2$.

At this stage, the PDF control problem has been formulated into the tracking problem for the above nonlinear weighting systems, and the control objective is to find $u(t)$ such that the tracking performance, disturbance attenuation performance and stability are guaranteed simultaneously.

For the transformed tracking control for the weights in $H_\infty$ setting, few tracking results have been provided for time delayed systems with nonlinearity (see [70, 202], or [63] and references therein for PI or PID tracking control problems). In this part, a new mixed $H_2$ and $H_\infty$ optimization formulation will be applied to the above weight tracking problem.

Denote $\mathrm{sym}(A) := A + A^T$ and $MM^T := \int_{-d(0)}^{0} \phi^T(\tau)\phi(\tau)\mathrm{d}\tau$. For the time delayed system with non-zero initial conditions, the following result can be obtained by applying the approach in [70, 71] to the case of truncated norms.

**Theorem 8.1** For $\Sigma_f$ with scalars $\gamma > 0$, $\mu$ and the weighting matrix $G$, if the optimization problem $\delta_1 := \min\{\alpha + \mathrm{tr}(M^T GM)\}$ subject to

$$\Phi = \begin{bmatrix} \Phi_{11} & \Phi_{12} \\ \Phi_{21} & \Phi_{22} \end{bmatrix} < 0, \qquad \begin{bmatrix} \alpha & \phi^T(0) \\ \phi(0) & Q \end{bmatrix} \geq 0 \tag{8.10}$$

is feasible with respect to $Q > 0$ and $L$, where

$$\Phi_{11} := \begin{bmatrix} \mathrm{sym}(AQ + B_2L) + G + \mu^2 FF^T & A_dQ + B_{2d}L \\ (A_dQ + B_{2d}L)^T & -(1-\beta)G \end{bmatrix},$$

$$\Phi_{12} := \begin{bmatrix} B_1 & QC_1^T & QC_0^T + L^T D_0^T & QU^T \\ 0 & 0 & 0 & 0 \end{bmatrix}, \quad \Phi_{22} := \begin{bmatrix} -\gamma^2 I & D_1^T & 0 & 0 \\ D_1 & -I & 0 & 0 \\ 0 & 0 & -I & 0 \\ 0 & 0 & 0 & -\mu^2 I \end{bmatrix},$$

then there exists a mixed guaranteed cost and generalized $H_\infty$ state feedback controller $u(t) = Kx(t)$, where $K = LQ^{-1}$, such that the closed-loop system is asymptotically stable and satisfies $J_2 \leq \delta(Q^{-1}, Q^{-1}GQ^{-1}) \leq \delta_1$ and $J_\infty(\gamma, P, S) < 0$.

*Proof.* Define a Lyapunov function condition as

$$\begin{aligned} V(x(t),t) := {} & x^T(t)Px(t) + \int_{t-d(t)}^{t} x^T(\tau)Sx(\tau)\mathrm{d}\tau \\ & + \int_0^t [\|\lambda Ux(\tau)\|^2 - \|\lambda f(x(\tau))\|^2]\mathrm{d}\tau. \end{aligned} \tag{8.11}$$

By using the Schur complement formula, and defining $\lambda = \mu^{-1}$, $\Phi < 0$ is equivalent to

$$\begin{bmatrix} \Psi & A_dQ + B_{2d}L & B_1 & QC_1^T & F \\ (A_dQ + B_{2d}L)^T & -(1-\beta)G & 0 & 0 & 0 \\ B_1^T & 0 & -\gamma^2 I & D_1^T & 0 \\ C_1Q & 0 & D_1 & -I & 0 \\ F^T & 0 & 0 & 0 & -\lambda^2 I \end{bmatrix} < 0 \tag{8.12}$$

where $\Psi = \mathrm{sym}(AQ + B_2L) + G + \lambda^2 QU^T UQ + (C_0Q + D_0L)^T (C_0Q + D_0L)$. From the third and fourth columns and rows of the left matrix in inequality (8.12) it can be seen that

$$\Phi' = \begin{bmatrix} \Psi & A_dQ+B_{2d}L & F \\ (A_dQ+B_{2d}L)^T & -(1-\beta)G & 0 \\ F^T & 0 & -\lambda^2 I \end{bmatrix} < 0. \quad (8.13)$$

Also, it is noted that

$$\begin{bmatrix} -\gamma^2 I & D_1^T \\ D_1 & -I \end{bmatrix} < 0$$

also holds, based on inequality (8.12) as a static condition.

Along with the solution of Equation 8.11 in the absence of $w(t)$, it can be shown that

$$\begin{aligned} \dot{V}(x(t),t) &= 2x^T(t)P\dot{x}(t) + x^T(t)Sx(t) - (1-\dot{d}(t))x^T(t-d)Sx(t-d) \\ &\quad + \|\lambda Ux(t)\|^2 - \|\lambda f(x(t))\|^2 \\ &\le x^T(t)(\mathrm{sym}(PA+PB_2K)+S)x(t) + x^T(t)(PA_d+PB_{2d}K)x(t-d) \\ &\quad + x^T(t-d)(A_d^TP+K^TB_{2d}^TP)x(t) - (1-\beta)x^T(t-d)Sx(t-d) \\ &\quad + x^T(t)PFf(x(t)) + f^T(x(t))F^TPx(t) + \|\lambda Ux(t)\|^2 - \|\lambda f(x(t))\|^2 \\ &= \varsigma^T(t)(\Omega - \Gamma\Gamma^T)\varsigma(t). \end{aligned}$$

$$\Omega = \begin{bmatrix} \Psi_1 & PA_d+PB_{2d}K & PF \\ (PA_d+PB_{2d}K)^T & -(1-\beta)S & 0 \\ F^TP & 0 & -\lambda^2 I \end{bmatrix} \quad (8.14)$$

where $\varsigma^T(t) = \begin{bmatrix} x^T(t) & x^T(t-d) & f^T(x(t)) \end{bmatrix}$, $\Gamma = \begin{bmatrix} (C_0+D_0K)^T & 0 & 0 \end{bmatrix}$, $\Psi_1 = P(A+B_2K)+(A+B_2K)^TP+S+\lambda^2U^TU+(C_0+D_0K)^T(C_0+D_0K)$. Pre-multiplying by $\mathrm{diag}\{P^{-1},I,I\}$ and post-multiplying by $\mathrm{diag}\{I,P^{-1},I\}$ on both side of inequality (8.12) and defining $Q=P^{-1}$, $K=LQ^{-1}$ and $G=QSQ$, it can be shown that inequality (8.14) is equivalent to inequality (8.13). Obversely $\Omega<0$ holds when $\Phi'<0$ holds. It can be seen that for any $\varsigma(t)\neq 0$, the following inequality holds:

$$\dot{V}(x(t)) \le \varsigma^T(t)(\Omega-\Gamma\Gamma^T)\varsigma(t) < -\varsigma^T(t)\Gamma\Gamma^T\varsigma(t) = -z_0^Tz_0 \le 0.$$

It follows that system (8.7) is asymptotically stable in the absence of the exogenous input.

Next we consider the performances to be optimized for system (8.7). Two auxiliary functions are selected as follows:

$$J_0 = z_0^Tz_0 + \dot{V}(x(t),t), \quad J_1 = z_\infty^T(t)z_\infty(t) - \gamma^2 w^T(t)w(t) + \dot{V}(x(t),t). \quad (8.15)$$

Following definition of the $H_2$ performance measure for time delay systems, we only consider the case in the absence of $w(t)$. By using the above formulation of $\dot{V}(x(t),t)$, it can be further verified that $J_0 \le \varsigma^T(t)\Omega\varsigma(t)$, where $\Omega$ is denoted as in Equation 8.14.

By using the Schur complement formula, and defining $\lambda = \mu^{-1}$, $\Phi < 0$ is equivalent to

$$\Phi'' = \begin{bmatrix} \Psi_2 & A_dQ + B_{2d}L & B_1 + QC_1^T D_1 & F \\ (A_dQ + B_{2d}L)^T & -(1-\beta)G & 0 & 0 \\ B_1^T + D_1^T C_1 Q & 0 & -\gamma^2 I + D_1^T D_1 & 0 \\ F^T & 0 & 0 & -\lambda^2 I \end{bmatrix} < 0 \tag{8.16}$$

where $\Psi_2 = \mathrm{sym}(AQ + B_2L) + G + \lambda^2 QU^T UQ + QC_1^T C_1 Q + (C_0Q + D_0L)^T (C_0Q + D_0L)$. In the presence of $w(t)$, taking the derivative of the function $V(x(t),t)$ along with the solution of system (8.7) yields that

$$\begin{aligned}\dot{V}(x(t),t) \le\; & x^T(t)(\mathrm{sym}(PA + PB_2K))x(t) + x^T(t)(PA_d + PB_{2d}K)x(t-d) \\ & + x^T(t-d)(A_d^T P + K^T B_{2d}^T P)x(t) + x^T(t)PB_1 w(t) + w^T(t)B_1^T Px(t) \\ & + x^T(t)Sx(t) - (1-\beta)x^T(t-d)Sx(t-d) + x^T(t)PFf(x(t)) \\ & + f^T(x(t))F^T Px(t) + \|\lambda Ux(t)\|^2 - \|\lambda f(x(t))\|^2\end{aligned}$$

with which it can be verified that $J_1 = \theta^T(t)(\Pi - \Gamma_1\Gamma_1^T)\theta(t)$,

$$\Pi = \begin{bmatrix} \Psi_3 & PA_d + PB_{2d}K & PB_1 + C_1^T D_1 & PF \\ (PA_d + PB_{2d}K)^T & -(1-\beta)S & 0 & 0 \\ B_1^T P + D_1^T C_1 & 0 & -\gamma^2 I + D_1^T D_1 & 0 \\ F^T P & 0 & 0 & -\lambda^2 I \end{bmatrix} \tag{8.17}$$

where $\theta^T(t) = \left[x^T(t)\; x^T(t-d)\; w^T(t)\; f^T(x(t))\right]$, $\Gamma_1 = \left[(C_0 + D_0K)^T\; 0\; 0\; 0\right]$, $\Psi_3 = \Psi_1 + C_1^T C_1$. Then pre-multiplying and post-multiplying both sides of inequality (8.17) by $\mathrm{diag}\{P^{-1}, I, I, I\}$ and $\mathrm{diag}\{I, P^{-1}, I, I\}$ respectively, we can show that inequality (8.17) is equivalent to inequality (8.16), which implies that $\Pi < 0$ since $\Phi'' < 0$. Thus it can be claimed that both $J_0 < 0$ and $J_1 < 0$ can be guaranteed under the condition $\Phi < 0$.

Based on Equations 8.5 and 8.11, it is shown that

$$\begin{aligned} V(x(T)) := & x^T(T)Px(T) + \int_{T-d(t)}^{T} x^T(\tau)Sx(\tau)d\tau \\ & + \int_0^T [\|\lambda Ux(\tau)\|^2 - \|\lambda f(x(\tau))\|^2]d\tau \ge 0. \end{aligned} \tag{8.18}$$

Integrating $J_0$ and $J_1$ from the initial time to $T$ implies that

$$0 \ge \int_0^T J_0(\tau)d\tau \ge \|z_0(t)\|_{2T}^2 + V(x(T)) - \delta(P,S) \ge \|z_0(t)\|_{2T}^2 - \delta(P,S), \tag{8.19}$$

$$0 \ge \int_0^T J_1(\tau)d\tau \ge \|z_\infty(t)\|_{2T}^2 - \gamma^2\|w\|_{2T}^2 + V(x(T)) - \delta(P,S)$$

$$\geq \|z_\infty(t)\|_{2T}^2 - \gamma^2\|w\|_{2T}^2 - \delta(P,S). \tag{8.20}$$

Thus, from inequalities (8.19) and (8.20), it can be obtained that $\|z_0(t)\|_{2T}^2 < \delta(P,S)$ and $J_\infty(\gamma,P,S) < 0$. □

*Remark 8.3* For the control problem, the mix $H_2/H_\infty$ performance has been addressed for linear or uncertain time delay systems [8]. In most proposed approaches for time delay systems, zero initial conditions are assumed for the standard $H_\infty$ control. However, non-zero initial conditions play an important role in many practical systems. Furthermore, zero initial conditions for time delay systems will make the upper bound of the $H_2$ performance measure zero. In this chapter the generalized $H_\infty$ performance is introduced for non-zero initial conditions so that mixed $H_2$ and $H_\infty$ performance measures can be performed simultaneously for time delay systems.

*Remark 8.4* The above result shows that the design procedures can be reduced to algorithms of a class of LMIs with respect to $Q$, $G$ and $L$. The design procedure can be described as follows:

1. For the target PDF, find the appropriate statistical information (such as mean, moment and entropy) to characterize its main stochastic behaviors and determine the performance index.
2. Use a B-spline expansion technique to model the measurable statistical information, with which the performance function $\delta(y,u(t))$ can be represented by the weight dynamics corresponding to some fixed basis functions.
3. Establish a further dynamical model between the input and the dynamic weight vectors. It is noted that the infinite dimensional STC can be reduced to a finite dimension weight tracking problem and more feasible design algorithms can be applied.
4. Calculate the controller gain based on LMIs to achieve the multiple objectives including stability, tracking performance, and mixed $H_2/H_\infty$ performance objectives.

## 8.4 Illustrative Examples

PDF shaping control for paper making has been investigated in [157]. In the head box, fibers, fillers and other chemical additives are mixed. This mixture generally consists of solids and water. When this mixture is injected onto the moving wire table, some water is drained through the wire nets into a white water pit underneath the wire table. This drainage process is continuous and the water in the pit also contains some solids in the form of either flocculation, or particles with a size distribution. As such, in order to control the efficiency of raw material usage, the total solids in the drained water need to be controlled and minimized. As discussed in [161], a group of chemicals, known as retention aids, can help to control this solid distribution. As a result, flocculation will occur and the density of the solid distribution can be locally increased. It can thus be concluded that these retention polymers are used to control (minimize) the size distribution of solid flocculation in the drained water. Indeed, since the head box can be regarded as a tank level system, at least a first order dynamic exists between the input (i.e., the retention aids) and the output distribution (i.e., the flocculation size distribution in the white water pit). With reference to several laboratory tests on the flocculation size distribution in the white water system, 100 samples have been formulated and used to evaluate 100 different PDFs. The distribution of flocculation sizes can be approximated by a truncated $\Gamma$-distribution under certain assumptions (see [139] for details).

Following the models given in [161], we consider a generalized model with additional perturbations, uncertainties and time delays (see Equations 8.2 and 8.4). It is supposed that function $\delta(t)$ can be formulated to be Equation 8.2 with $V(t) = [v_1 \ \ v_2 \ \ v_3 \ \ v_4 \ \ v_5 \ \ v_6 \ \ v_7 \ \ v_8 \ \ v_9]^T$ and

$$B_i(y) = \mathrm{e}^{(-(y-y_i)^2\sigma_i^{-2})} \quad (i = 1,2,\cdots,9)$$

where $y \in [0, 0.05]$, $y_i = 0.003 + 0.006(i-1)$, $\sigma_i = 0.003$ $(i = 1,2,\cdots,9)$ after a B-spline approximation procedure. For simplicity, in simulation it is assumed that the target PDF can be denoted as follows

$$\gamma_g(y) = \begin{cases} \frac{\lambda^r}{\Gamma(r)} y^{r-1}\mathrm{e}^{-\lambda y}, & when \ \ y > 0 \\ 0, & when \ \ y \leq 0 \end{cases}$$

where $r = 2$ and $\lambda = 1$. From Equation 8.1, we can obtain

$$\Sigma_{i=1}^{9}(V_{gi}B_i(y)) = Q_1\gamma_g(y)ln(\gamma_g(y)) + Q_2 y\gamma_g(y)$$

where $Q_1 = 1$, $Q_2 = 1$. Therefore we have $\Sigma_{i=1}^{9}(V_{gi}\int_{-\infty}^{\infty}(B_i(y)\mathrm{d}y)) = 0.422784$. Because $\int_{-\infty}^{\infty} B_i(y)\mathrm{d}y = \sqrt{\pi}\sigma_i$, it can be shown that the reference weights corresponding to the target statistical information of the target PDF satisfy the condition $\Sigma_{i=1}^{9}(V_{gi}) = 15.9$. As such the reference weights can be denoted $V_g = [1 \ \ 2 \ \ 2 \ \ 1 \ \ 2 \ \ 3 \ \ 1.5 \ \ 2 \ \ 2.4]^T$.

The dynamic model between $u$ and $x$ is described by Equation 8.7. Similarly to the modeling procedures of [15], the coefficient matrices are denoted by

$$A = \mathrm{diag}\{-0.83, -0.83, -0.83, -0.83, -0.83, -0.83, -0.83, -0.83, -0.83\},$$

$$A_d = \mathrm{diag}\{-0.2, -0.2, -0.2, -0.2, -0.2, -0.2, -0.2, -0.2, -0.2\},$$

$$B_{12} = \mathrm{diag}\{-0.2, -0.2, -0.2, -0.2, -0.2, -0.2, -0.2, -0.2, -0.2\},$$

$$B_{2d} = \mathrm{diag}\{-0.2, -0.2, -0.2, -0.2, -0.2, -0.2, -0.2, -0.2, -0.2\},$$

$$B_2 = \mathrm{diag}\{0.1, 0.1, 0.1, 0.1, 0.1, 0.1, 0.1, 0.1, 0.1\},$$

$$F = [-1 \ \ 1 \ \ 0.5 \ \ -0.5 \ \ 1 \ \ 0.5 \ \ 0 \ \ -0.5 \ \ 1.2]^T,$$

$$B_{11} = [0.2 \ \ 0.2 \ \ 0.2 \ \ 0.2 \ \ 0.2 \ \ 0.2 \ \ 0.2 \ \ 0.2 \ \ 0.2]^T,$$

$$C_0 = [1.2\ \ -1.2\ \ -0.5\ \ 1\ \ 0\ \ 1\ \ -1\ \ 0.2\ \ -1],$$

$$C_1 = [-0.2\ \ -0.2\ \ -0.2\ \ -0.2\ \ -0.2\ \ -0.2\ \ -0.2\ \ -0.2\ \ -0.2],$$

$$D_0 = [0.5\ \ -1\ \ 0.6\ \ 0.1\ \ -1\ \ 0\ \ 0.2\ \ 1\ \ -1],$$

$$D_1 = [0.01\ \ 0.1\ \ 0.1\ \ -0.1\ \ 0.5\ \ 1\ \ 0\ \ -1\ \ 0\ \ -1.2],$$

$$U = [-0.1\ \ -0.2\ \ 1\ \ -1\ \ -0.1\ \ -0.2\ \ 1\ \ -1\ \ -0.5].$$

Through solving the LMI (8.10), the matrices can be computed as follows:

$$Q = \begin{bmatrix} 6.93 & -0.30 & -0.39 & -0.44 & -0.34 & -0.53 & -0.13 & -0.64 & 0.21 \\ -0.30 & 6.70 & 0.11 & -0.38 & -0.40 & -0.09 & -0.22 & -0.07 & -1.11 \\ -0.38 & 0.11 & 5.33 & 0.95 & -0.06 & -0.13 & -1.78 & 0.81 & 0.49 \\ -0.44 & -0.38 & 0.95 & 5.66 & -0.27 & -0.52 & 1.03 & -1.61 & -0.74 \\ -0.33 & -0.40 & -0.06 & -0.27 & 6.84 & -0.09 & -0.36 & -0.02 & -0.97 \\ -0.52 & -0.09 & -0.13 & -0.52 & -0.08 & 6.93 & -0.19 & -0.56 & -0.49 \\ -0.12 & -0.22 & -1.78 & 1.00 & -0.36 & -0.19 & 5.49 & 0.98 & 0.48 \\ -0.64 & -0.08 & 0.82 & -1.61 & -0.02 & -0.56 & 0.98 & 5.30 & -0.49 \\ 0.22 & -1.11 & 0.49 & -0.74 & -0.97 & -0.49 & 0.48 & -0.48 & 6.55 \end{bmatrix},$$

$$L = \begin{bmatrix} 37.19 & 7.56 & -2.52 & -5.08 & 7.46 & -0.16 & -2.17 & -9.92 & -1.12 \\ 7.53 & 29.87 & 2.16 & 1.19 & -15.78 & -3.79 & 0.34 & 9.79 & -6.64 \\ -3.16 & 3.45 & 32.81 & 4.37 & 2.69 & -1.78 & -9.96 & -0.77 & 8.63 \\ -5.27 & 1.53 & 4.41 & 37.25 & 2.02 & -1.54 & 4.03 & -10.17 & -5.19 \\ 6.48 & -14.02 & 0.53 & 1.44 & 31.93 & -3.75 & -0.60 & 8.52 & -4.71 \\ -1.63 & -1.08 & -3.13 & -1.96 & -1.12 & 43.21 & -1.57 & -3.78 & 1.72 \\ -1.69 & -0.46 & -9.41 & 4.22 & -1.10 & -1.18 & 37.03 & 3.38 & 2.33 \\ -9.79 & 9.54 & 0.65 & -9.85 & 9.98 & -1.18 & 2.70 & 28.11 & 0.89 \\ 1.37 & 11.32 & 9.77 & -4.72 & -10.81 & -0.65 & 3.63 & 5.64 & 27.58 \end{bmatrix},$$

$$K = \begin{bmatrix} 5.33 & 1.35 & 0.55 & -1.01 & 1.39 & 0.21 & 0.56 & -1.69 & -0.22 \\ 1.49 & 4.48 & -0.55 & 1.52 & -1.96 & -0.10 & -0.79 & 2.76 & -0.12 \\ -0.25 & 0.50 & 6.70 & -0.81 & 0.57 & -0.25 & 0.76 & -1.55 & 0.71 \\ -0.22 & 0.74 & -0.96 & 7.28 & 0.63 & 0.37 & -0.99 & 0.69 & 0.48 \\ 1.34 & -1.65 & -0.57 & 1.38 & 4.66 & -0.12 & -0.72 & 2.40 & 0.06 \\ 0.28 & 0.17 & -1.04 & 0.93 & 0.08 & 6.42 & -0.77 & 0.69 & 1.07 \\ -0.25 & -0.04 & 1.34 & -1.41 & 0.09 & -0.22 & 7.76 & -1.54 & -0.59 \\ -0.74 & 1.86 & -1.61 & 0.90 & 1.73 & 0.43 & -1.25 & 6.18 & 1.54 \\ 0.13 & -1.17 & 1.79 & -0.67 & -1.09 & 0.22 & 0.79 & 0.79 & 3.65 \end{bmatrix},$$

$$G = \begin{bmatrix} -13.99 & 4.05 & 0.69 & 3.18 & 4.33 & 0.19 & 0.70 & -0.41 & 3.91 \\ -3.69 & -13.50 & 3.48 & -4.62 & -5.93 & -3.99 & -1.04 & -15.91 & -7.91 \\ 0.49 & -5.12 & -14.15 & 4.75 & 3.70 & 0.01 & 9.76 & 22.50 & -13.57 \\ -4.28 & 5.03 & -3.25 & -14.27 & 3.73 & 1.12 & -8.07 & 7.55 & 0.27 \\ -3.88 & 5.64 & -5.40 & -3.21 & -13.65 & 1.23 & 3.94 & -1.73 & -12.58 \\ 0.22 & 3.12 & -0.39 & -1.02 & -2.12 & -13.67 & -0.27 & -0.21 & 3.69 \\ -0.39 & 0.50 & -11.06 & 9.06 & -4.56 & 0.12 & -14.04 & 0.92 & 5.01 \\ -0.19 & 15.37 & -20.42 & -9.10 & 1.41 & 0.44 & 0.19 & -13.77 & -15.06 \\ -6.42 & 10.50 & 14.25 & -2.21 & 15.16 & -3.36 & -4.54 & 12.02 & -13.94 \end{bmatrix}.$$

When the control input $u = Kx(t)$ is applied, the closed-loop system responses for the dynamic weight are showed in Figures 8.1-8.3. These three figures represent the nine weights in terms of tracking errors. As expected, they all converge to zero and thus show that the control strategy can stabilize the system. Figures 8.4-8.6 display the 3D mesh plot of statistical function $\delta(y,t)$, where it can be seen that this statistical function has been controlled to follow its target distribution as defined by $V_g$.

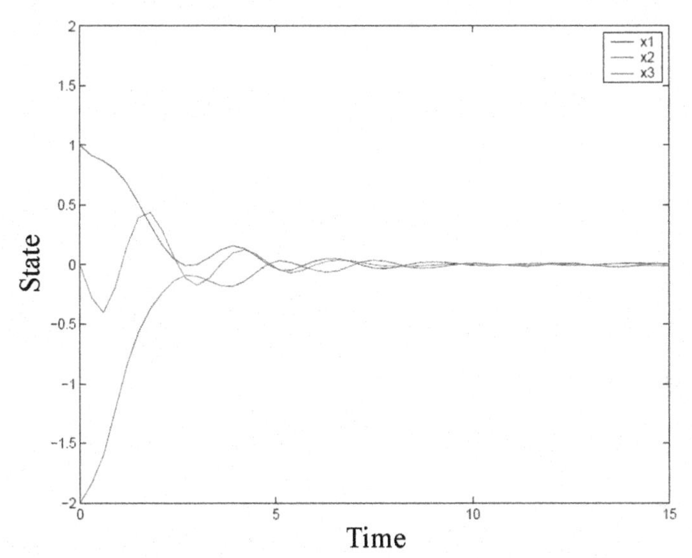


**Figure 8.1** Responses of dynamic weight vectors

## 8.5 Conclusions

Different from any other stochastic tracking problems, this chapter presents a new stochastic control setting where the tracking objective is presented as some statistical information rather than a deterministic reference signal or a PDF. It is called the STC problem. B-spline NN are used to approximate the integrated function related to the tracked mean or entropy. By establishing the weight systems, we have introduced the generalized $\mathrm{H}_2/H_\infty$ optimization problem to formulate the weight tracking. LMI-based approaches are proposed to fulfill the multi-objective optimal controller design where the initial condition are non-zero and the truncated norm is adopted. Stability analysis for tracking control is developed by Lyapunov stability criteria. Simulations are given to demonstrate the efficiency of the proposed approach.

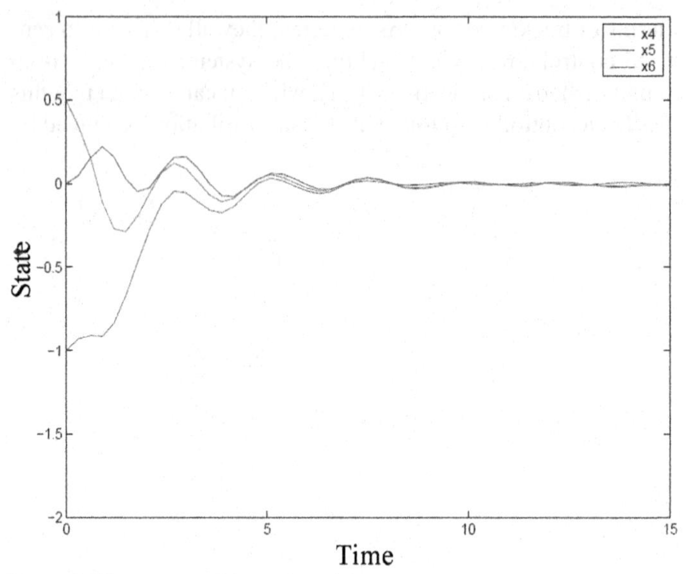


**Figure 8.2** Responses of dynamic weight vectors

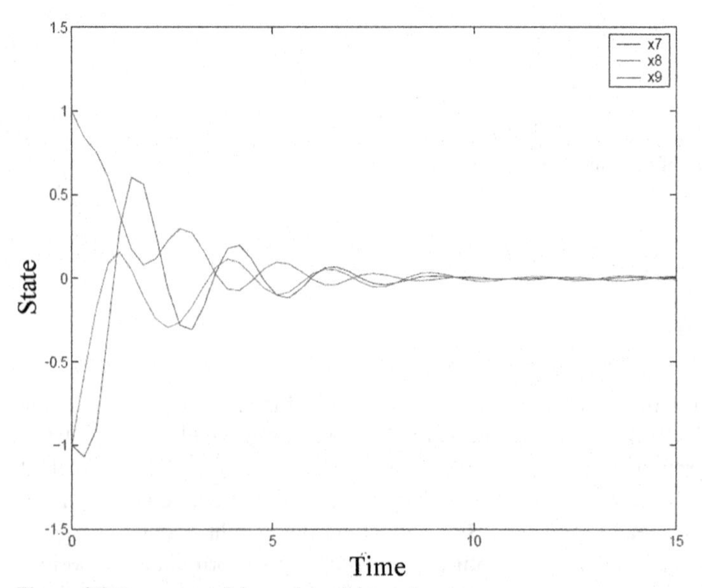


**Figure 8.3** Responses of dynamic weight vectors

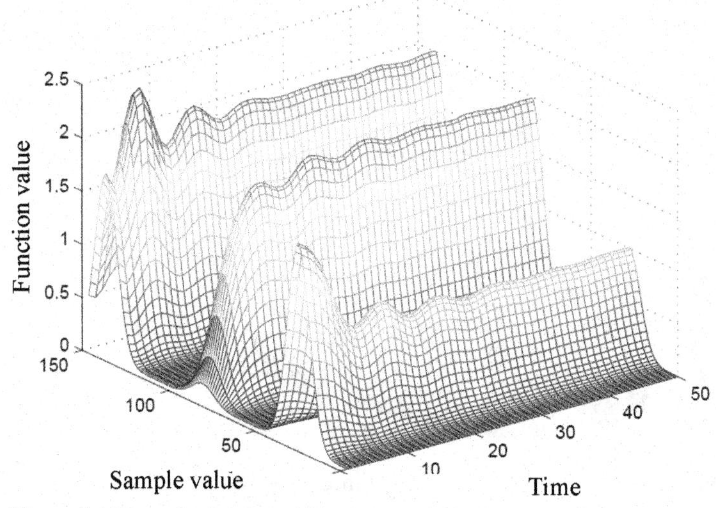


**Figure 8.4** 3D mesh of statistical function

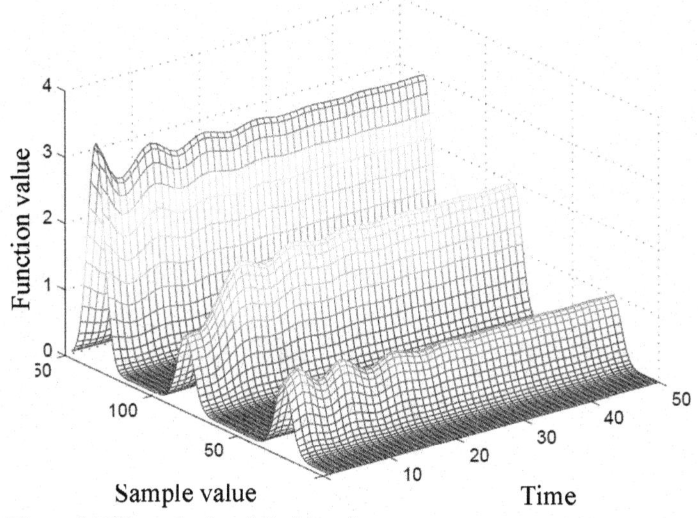


**Figure 8.5** 3D mesh of statistical function

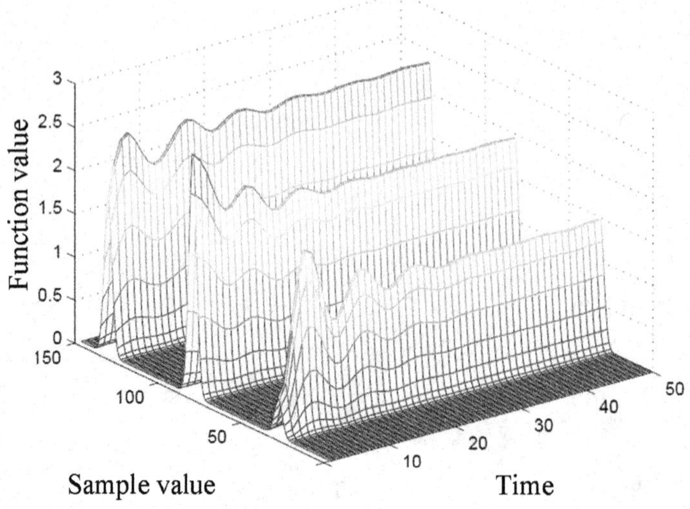


**Figure 8.6** 3D mesh of statistical function

# Chapter 9
# Adaptive Statistical Tracking Control Based on Two-step Neural Networks with Time Delays

## 9.1 Introduction

Motivated by Chapter 6 and Chapter 8, in this chapter two-step NN are applied to solve the STC problem. Similarly to the output PDF tracking control problem, after using B-spline NN approximation theory for the integrated performance function in the context of the entropy and mean, it is shown that the above STC problem can be transformed into a tracking problem for the weighting dynamics. However, due to lack of model knowledge, most published results only concern linear and precise models and some nonlinear models discussed were also difficult to obtain through traditional identification approaches. In this chapter, we attempt to import time delay DNNs with undetermined parameters to identify the unknown nonlinear dynamic relationships. This represents a significant extension to the previous results. In the proposed method, the adaptive laws and LMI obtained can guarantee that the identification error converges to zero within a finite time. Based on adaptive projection arithmetic (see [170, 171]) and VSC theory (see [21, 104, 152]), a dynamic state feedback controller is developed in which the weight vectors can be made to follow the outputs of a reference model with a fast response. Finally, simulations are provided to show the effectiveness of the proposed results.

## 9.2 Problem Formulation and Preliminaries

For a dynamical stochastic system, denote $u(t) \in R^m$ as the control input, $z(t) \in [a,b]$ as the stochastic output, whose conditional PDF is denoted by $\gamma(y,u(t))$, where $y$ is the variable in the sample interval $[\alpha,\beta]$. It is noted that $\gamma(y,u(t))$ is a dynamic functional of $y$ along with the time variable $t$. Similarly to Chapter 8, in this part we consider a STC problem. The same performance index is $\int_\alpha^\beta \delta(\gamma,u(t))\mathrm{d}y$, where

$$\delta(\gamma,u(t)) = Q_1\gamma(y,u(t))\ln(\gamma(y,u(t))) + Q_2 y\gamma(y,u(t)) \tag{9.1}$$

where $Q_1$ and $Q_2$ are two designed parameters. In Equation 9.1 the integral of the first term is the entropy and that of the second term is the mean of the output variable.

In the following, we construct B-spline NN to approximate $\delta(\gamma, u(t))$ as follows:

$$\delta(\gamma, u(t)) = C(y)V(t) \tag{9.2}$$

where

$$C(y) = [c_1(y), \dots c_n(y)], \quad V(t) = [v_1(t), \dots v_n(t)]^T.$$

$c_i(y)$ $(i = 1, 2, \cdots, n)$ are pre-specified basis functions and $v_i(u(t)) := v_i(t), (i = 1, 2, \cdots, n)$ are the corresponding weights. Based on Equations 9.1 and 9.2, for the target PDF we can compute the statistical information. Furthermore, we can find the corresponding weights, which can be denoted $V_g(t)$. That is, $\delta(g, u(t)) = C(y)V_g(t)$. The tracking objective is to find $u(t)$ such that $\delta(\gamma, u(t)) - \delta(g, u(t)) = C(y)W(t)$ converges to 0, where $W(t) := V(t) - V_g(t)$.

## 9.3 Time Delay DNNs Identification

Once B-spline expansions have been made for the integrated function (9.1), the next step is to find the dynamic relationships between the control input and the weight vectors $V(t)$ corresponding to a further modeling procedure. It is well known that DNN identifiers can be employed to perform black box identification. Similarly to Chapter 6, we will provide a time delay DNN model to characterize the nonlinear dynamics, with a learning strategy for the model parameters. Then a novel adaptive variable structure tracking controller will be given for tracking the target weights.

It is assumed that there exists a set of optimal model parameters (which can also be seen as weight matrices related to the DNNs) $W_1^*$, $W_2^*$ such that the unknown nonlinear dynamics between the control input $u(t)$ and the weight vectors $x(t) := V(t)$ can be described by the following NN model:

$$\begin{aligned} \dot{x}(t) &= Ax(t) + A_1 x_\tau(t) + BW_1^* \sigma(x_\tau(t)) \\ &\quad + BW_2^* \phi(x(t))u(t) - DF(t) \end{aligned} \tag{9.3}$$

where $x(t) \in R^n$ is the measurable state of the unknown dynamic model and $x_\tau(t) := x(t - \tau(t))$ is the state with a time delay term. $A$ and $A_1 \in R^{n \times n}$ are stable matrices, $B$ and $D$ are known coefficient matrices with compatible dimensions. $F(t)$ represents the error term. Obviously the optimal model parameters $W_1^*$, $W_2^*$ are bounded unknown matrices and satisfy $W_1^* W_1^{*T} \leq \bar{W}_1$, $W_2^* W_2^{*T} \leq \bar{W}_2$, where $\bar{W}_1$ and $\bar{W}_2$ are prior known positive definite matrices (see [189] for details).

To formulate the required algorithm, the following assumptions are made.

**Assumption 9.1** *The time delay term $\tau(t)$ is continuous and is assumed to satisfy $\dot{\tau}(t) \leq \beta < 1$, where $0 < \beta < 1$ is a known positive parameter.*

**Assumption 9.2** *There exists an unknown positive constant d such that the error term satisfies* $\|F(t)\|_1 \leq d$.

**Assumption 9.3** *The control input* $u(t)$ *is assumed bounded and satisfy* $u^T(t)u(t) \leq \bar{u}$, *where* $\bar{u}$ *is a known positive constant.*

Vector functions $\sigma(x) \in R^m$ are assumed to be $m$-dimensional with elements increasing monotonically, and matrix function $\phi(x)$ is assumed to be an $m \times m$ diagonal matrix. In this context, typical presentation of the elements $\sigma_i(.), \phi_i(.)$ are as sigmoid functions, i.e.,

$$\sigma_i(x_i) = \frac{a}{1+\mathrm{e}^{-bx_i}} - c. \tag{9.4}$$

It can be seen that the Hopfield NN model is a special case of this kind of NN with $A = \mathrm{diag}\{-\frac{1}{R_iC_i}\}$, where $R_i$ and $C_i$ represent the resistance and capacitance,respectively, at the ith node of the network.

We construct the following DNNs for system identification:

$$\dot{\hat{x}}(t) = A\hat{x}(t) + A_1\hat{x}_\tau(t) + BW_1\sigma(\hat{x}_\tau(t)) + BW_2\phi(\hat{x}(t))u(t) + u_f(t) \tag{9.5}$$

where $\hat{x}(t) \in R^n$ is the state of the DNN, the initial condition is assumed to be $\hat{x}(t) = 0, t \in [-\tau(t), 0]$. $W_1 \in R^{m\times m}$ are the synaptic weights of DNNs and $W_2$ is a $m \times m$ diagonal matrix of the form $W_2 = \mathrm{diag}[w_{21}, w_{22}, \cdots w_{2m}]$. It is noted that the optimal parameters $W_1^*$, $W_2^*$ can be seen as optimal estimation values of the weights $W_1$ and $W_2$, implying $W_2^*$ is a diagonal matrix. $u_f(t)$ is the compensation term and will be defined later.

Denoting $\tilde{W}_1 = W_1 - W_1^*, \tilde{W}_2 = W_1 - W_2^*$, $\tilde{\sigma} = \sigma(\hat{x}_\tau(t)) - \sigma(x_\tau(t)), \tilde{\phi} = \phi(\hat{x}(t)) - \phi(x(t))$ and the identification error as $e(t) = \hat{x}(t) - x(t)$. Because $\sigma(.)$ and $\phi(.)$ are chosen as sigmoid functions, they satisfy the following Lipschitz property

$$\tilde{\sigma}^T\tilde{\sigma} \leq \mathrm{e}^T(t-\tau(t))E_\sigma e(t-\tau(t)), \;\; (\tilde{\phi}u(t))^T(\tilde{\phi}u(t)) \leq \bar{u}\mathrm{e}^T(t)E_\phi e(t) \tag{9.6}$$

where $E_\sigma, E_\phi$ are known positive definite matrices.

From Equations 9.3 and 9.5, the identification error satisfies

$$\dot{e}(t) = Ae(t) + A_1e(t-\tau(t)) + B\tilde{W}_1\sigma(\hat{x}_\tau) + B\tilde{W}_2\phi(\hat{x})u(t)$$
$$+BW_1^*\tilde{\sigma} + BW_2^*\tilde{\phi}u(t) + u_f(t) + DF(t). \tag{9.7}$$

In the following, we will consider the stable learning procedure of DNN identification with a smooth time delay.

**Theorem 9.1** Consider the identification scheme (9.7); if the following LMI

$$\Phi = \begin{bmatrix} \Psi & PA_1 & PB \\ A_1^TP & -(1-\beta)U + E_\sigma & 0 \\ B^TP & 0 & -R \end{bmatrix} < 0 \tag{9.8}$$

is solvable for $P > 0$, $Q_0 > 0$, $U > 0$ where $\Psi = PA + A^T P + U + \bar{u}E_\phi + Q_0$, $R = (\bar{W}_1 + \bar{W}_2)^{-1} > 0$. The compensation term is given as

$$u_f(t) = -D \cdot SGN\{D^T Pe(t)\}\hat{d}. \tag{9.9}$$

The model parameters $W_1$ , $W_2$ and $\hat{d}$ are updated as

$$\begin{aligned} \dot{W}_1 &= -\gamma_1 B^T Pe(t)\sigma^T(\hat{x}_\tau), \\ \dot{W}_2 &= -\gamma_2 E[B^T Pe(t)u^T \phi(\hat{x})], \\ \dot{\hat{d}} &= \gamma_3 \|D^T Pe(t)\|_1. \end{aligned} \tag{9.10}$$

$\hat{d}$ is an estimation value of unknown constant $d$, and $\gamma_i$, $i = 1, 2, 3$, are positive defined constants. $E[.]$ represents a kind of matrix transformation that makes a common matrix into a diagonal form. Then it can be obtained that $\lim_{t\to\infty} e(t) = 0$.

*Proof.* Similar to the proof of Theorem 6.1. The proof process is omitted. □

*Remark* 9.1 The PDF $\gamma(y, u(t))$, as the control objective, can be measured or estimated using instruments (such as a laser particle size distribution measure or a digital camera) or the kernel estimation technique based on an open-loop test. So the statistical performance function $\delta(y, u(t))$ is also measurable. Meanwhile, because $C(y)$ is the pre-specified basis function vector, which is related to the B-spline NN, the weight vectors $V(t)$ are also measurable. Therefore the errors $e(t)$, adaptive law and compensation term $u_f(t)$ can be obtained.

## 9.4 Variable Structure Tracking Control for Weight Vectors

In this section we investigate the tracking controller design for DNN (9.5). A dynamic reference model is considered here as part of the target model whose output $x_m(t)$ converges to $V_g(t)$ as $t \to \infty$. Indeed, the dynamic reference model has advantages in adjusting the closed-loop dynamic transient behavior and has been used widely in the past for model following control and model reference adaptive control. It is supposed that the reference vector has the following dynamics:

$$\dot{x}_m(t) = A_m x_m(t) + B_m r \tag{9.11}$$

where $x_m \in R^n$ is the state vector, $A_m, B_m$ are constant matrices of appropriate dimensions. $r$ represents the control input. In order to guarantee $\lim_{t\to\infty} x_m(t) = V_g(t)$, the following condition for parameter matrices should be satisfied (see [14] for details)

$$V_g = (I - A_m)^{-1} B_m r.$$

At this stage, the problem is reduced to a nonlinear dynamical tracking control for error vector $e_1(t) = \hat{x}(t) - x_m(t)$. From Equations 9.5 and 9.11, we obtain

$$\begin{aligned}\dot{e}_1(t) &= A\hat{x}(t) + A_1\hat{x}_\tau(t) + BW_1\sigma(\hat{x}_\tau(t)) + BW_2\phi(\hat{x}(t))u(t)\\ &\quad -B_m r + u_f(t) - A_m x_m(t)\\ &= A_m e_1(t) + (A - A_m)\hat{x}(t) + A_1\hat{x}_\tau(t) + BW_1\sigma(\hat{x}_\tau(t))\\ &\quad +BW_2\phi(\hat{x}(t))u(t) - B_m r + u_f(t). \end{aligned} \tag{9.12}$$

In order to stabilize the tracking error dynamics (9.12), we introduce the VSC technique. In general, the VSC design procedure can be divided into two phases. First, it is necessary to design a sliding hyperplane so that the system dynamics will be driven into the desired sliding surface by using the VSC. The sliding function $S(t)$ is designed as (see [21])

$$S(t) = Ge_1(t) - \int_0^t G(A_m + BT)e_1(\alpha)\mathrm{d}\alpha \tag{9.13}$$

where $G \in R^{m\times n}$, $T \in R^{m\times n}$ are constant matrices and satisfy the following assumptions (see [21] for details).

**Assumption 9.4** *$G$ is chosen such that $GB$ is nonsingular.*

**Assumption 9.5** *Matrix $T$ is designed to satisfy the following inequality*

$$Re[\lambda_{max}(A_m + BT)] < 0. \tag{9.14}$$

**Assumption 9.6** Matrices $A - A_m$, $A_1$, $B_m$ and $D$ satisfy the following matching conditions in the sense that there exist matrices $M$, $H$, $N$ and $I$ of appropriate dimensions such that

$$A - A_m = BM, \quad A_1 = BN, \quad B_m = BH, \quad D = BI. \tag{9.15}$$

After designing the sliding hyperplane, the next phase of the traditional VSC design is to find an appropriate control law such that the sliding hyperplane will attract the trajectories of Equation 9.13, and these trajectories will remain on the sliding hyperplane for all subsequent time. In this context, we compute the time derivative of $S(t)$. Using Equations 9.12 and 9.13 we obtain

$$\begin{aligned}\dot{S}(t) &= G(A - A_m)\hat{x}(t) + GBW_1\sigma(\hat{x}_\tau(t)) + GBW_2\phi(\hat{x}(t))u(t)\\ &\quad +GA_1\hat{x}_\tau(t) - GB_m r + Gu_f(t) - GBTe_1(t). \end{aligned} \tag{9.16}$$

Taking $u(t)$ to be given by

$$u(t) = -[GBW_2\phi(\hat{x}(t))]^{-1}[G(A - A_m)\hat{x} + GA_1\hat{x}_\tau(t)$$

$$+GBW_1\sigma(\hat{x}_\tau(t)) - GB_m r + Gu_f(t) - GBTe_1(t) + \omega S(t)] \quad (9.17)$$

where $\omega$ is a designed positive constant and substituting $u(t)$ into Equation 9.16 we have

$$\dot{S}(t) = -\omega S(t).$$

In order to ensure the existence of $[GBW_2\phi(\hat{x}(t))]^{-1}$, we need to establish $w_{2i} \neq 0,\ i = 1, \cdots m$. Furthermore, from Equation 9.10 it can be seen that weight matrices $W_1$ and $W_2$ cannot be guaranteed within reasonable bounds. Hence $W_1$ and $W_2$ are re-confined through use of the projection algorithm. The standard adaptive laws are modified to read

$$\dot{W}_1 = \begin{cases} -\gamma_1 B^T Pe(t)\sigma^T(\hat{x}_\tau) \quad when \ \|W_1\| < M_1 \ or \ \|W_1\| = M_1 \\ \quad and \ \mathrm{tr}\{\sigma(\hat{x}_\tau)\mathrm{e}^T(t)PBW_1\} \geq 0 \\ -\gamma_1 B^T Pe(t)\sigma^T(\hat{x}_\tau) + \gamma_1 \mathrm{tr}\{\sigma(\hat{x}_\tau)\mathrm{e}^T(t)PBW_1\}\frac{W_1}{\|W_1\|^2} \\ \quad when \ \|W_1\| = M_1 \ and \ \mathrm{tr}\{\sigma(\hat{x}_\tau)\mathrm{e}^T(t)PBW_1\} < 0 \end{cases} \quad (9.18)$$

when $w_{2i} = \varepsilon$, we adopt

$$\dot{w}_{2i} = \begin{cases} -\gamma_2 u_i\phi_i(\hat{x})B_i^T Pe(t) \quad when \ u_i\phi_i(\hat{x})B_i^T Pe(t) < 0 \\ 0 \quad when \ u_i\phi_i(\hat{x})B_i^T Pe(t) \geq 0 \end{cases} \quad (9.19)$$

where $u_i$ is the ith element of $u(t)$ and $B_i$ is the ith column of $B$. Otherwise

$$\dot{W}_2 = \begin{cases} -\gamma_2 E[B^T Pe(t)u^T(t)\phi(\hat{x})] \ when \ \|W_2\| < M_2 \ or \ \|W_2\| = M_2 \\ \quad and \ \mathrm{tr}\{\phi(\hat{x})u(t)\mathrm{e}^T(t)PBW_2\} \geq 0 \\ -\gamma_2 E[B^T Pe(t)u^T(t)\phi(\hat{x})] + \gamma_2 tr\{\phi(\hat{x})u(t)\mathrm{e}^T(t)PBW_2\}\frac{W_2}{\|W_2\|^2} \\ \quad when \ \|W_2\| = M_2 \ and \ \mathrm{tr}\{\phi(\hat{x})u(t)e^T(t)PBW_2\} < 0 \ . \end{cases} \quad (9.20)$$

At this stage, the following theorem can be obtained.

**Theorem 9.2** Consider DNN model (9.5) and dynamic reference model (9.11), if the LMI (9.8) are solvable for $P > 0$, $U > 0$ then when control law (9.17) and the adaptive law (9.18-9.20) are applied, we have:

1. $\|W_1\| \leq M_1; \|W_2\| \leq M_2,\ w_{2i} \geq \varepsilon$, where $M_1, M_2$ and $\varepsilon$ are known constants and satisfy $M_1^2 > \mathrm{tr}(\bar{W}_1)$, $M_2^2 > \mathrm{tr}(\bar{W}_2)$.
2. $\lim_{t\to\infty} e(t) = \lim_{t\to\infty} S(t) = 0$, $\lim_{t\to\infty} e_1(t) = 0$.

*Proof.*

1. We take the Lyapunov function candidate $v = \frac{1}{2}\mathrm{tr}\{W_2^T\gamma_2^{-1}W_2\}$, then $\dot{v}$ can be expressed as

$$\dot{v} = -\mathrm{tr}\{\phi(\hat{x})u(t)\mathrm{e}^T(t)PBW_2\} + \bar{I}\mathrm{tr}\{\phi(\hat{x})u(t)\mathrm{e}^T(t)PBW_2\}\frac{\mathrm{tr}(W_2^T W_2)}{\|W_2\|^2}$$

where $\bar{I} = 0(1)$, if the first (second) condition of Equation 9.20 is true. If the first line of Equation 9.20 is true, $\dot{v} < 0$; if the second line of Equation 9.20 is true,

$\|W_2\| = M_2$, $\dot{v} = 0$, Therefore we have $\|W_2\| \leq M_2$. From Equation 9.19, if $w_{2i} = \varepsilon$, then $\dot{w}_{2i} \geq 0$, This means that $w_{2i} \geq \varepsilon$.

2. Select a Lyapunov function as

$$V_2(t) = \mathrm{e}^T(t)Pe(t) + \int_{t-\tau(t)}^{t} \mathrm{e}^T(\alpha)Ue(\alpha)\mathrm{d}\alpha + \frac{1}{2}S^T(t)S(t)$$

$$+\mathrm{tr}\{\tilde{W}_1^T\gamma_1^{-1}\tilde{W}_1\} + \mathrm{tr}\{\tilde{W}_2^T\gamma_2^{-1}\tilde{W}_2\} + \tilde{d}^T\gamma_3^{-1}\tilde{d}. \quad (9.21)$$

From Equations 9.12-9.17, $\dot{V}_2(t)$ can be expressed as

$$\dot{V}_2(t) \leq -\mathrm{e}^T(t)Q_0e(t) - \omega S^T(t)S(t) + 2\bar{I}_1\mathrm{tr}\{\sigma(\hat{x})e^T(t)PBW_1\}\mathrm{tr}\{\frac{W_1^T\tilde{W}_1}{\|W_1\|^2}\}$$

$$+2\bar{I}_2\mathrm{tr}\{\phi(\hat{x})u(t)\mathrm{e}^T(t)PBW_2\}\mathrm{tr}\{\frac{W_2^T\tilde{W}_2}{\|W_2\|^2}\}$$

where $\bar{I}_1 = 0(1)$, if the first (second) condition of Equation 9.18 is true. $\bar{I}_2 = 0(1)$, if the first (second) condition of Equation 9.20 is true. When the second condition of Equations 9.18 and 9.20 are true, we have $\|W_i\| = M_i > \sqrt{\mathrm{tr}(\bar{W}_i)} \geq \|W_i^*\|, i = 1,2$. Therefore we can get

$$\mathrm{tr}\{W_i^T\tilde{W}_i\} = \frac{1}{2}\mathrm{tr}[W_i^TW_i - W_1^{*^T}W_i^* + \tilde{W}_i^T\tilde{W}_i] \geq 0. \quad (9.22)$$

It can be seen that $\bar{I}_1\mathrm{tr}\{\sigma(\hat{x})\mathrm{e}^T(t)PBW_1\}\mathrm{tr}\{\frac{W_1^T\tilde{W}_1}{\|W_1\|^2}\} \leq 0$, Similarly, we have

$$\bar{I}_2\mathrm{tr}\{\mathrm{tr}\{\phi(\hat{x})u(t)\mathrm{e}^T(t)PBW_2\}\frac{W_2^T\tilde{W}_2}{\|W_2\|^2} \leq 0$$

. As a result, it can be shown that

$$\dot{V}_2(t) \leq -\mathrm{e}^T(t)Q_0e(t) - \omega S^T(t)S(t) \leq 0. \quad (9.23)$$

Similarly to Theorem 9.1, using the well-known Barbalat lemma, it can be obtained that $\lim_{t\to\infty}e(t) = \lim_{t\to\infty}S(t) = 0$.

Substituting Equation 9.17 into Equation 9.12, it can be obtained that

$$\dot{e}_1(t) = A_me_1(t) + BM\hat{x}(t) + BN\hat{x}_\tau(t) + BW_1\sigma(\hat{x}_\tau(t))$$

$$+BW_2\phi(\hat{x}(t))u(t) - BHr - BI\cdot SGN\{D^TPe\}\hat{d}. \quad (9.24)$$

The fact that $S(t)$ will approach zero means that the system dynamics will be driven into the desired sliding manifold. When the system is in the sliding mode it is known that $S(t) = 0$ and $\dot{S}(t) = 0$; it is also known that there exists an equivalent controller

$$u_{eq}(t) = -[GBW_2\phi(\hat{x}(t))]^{-1}GB[M\hat{x}(t) + N\hat{x}_\tau(t)$$

$$+W_1\sigma(\hat{x}_\tau(t)) - I\cdot SGN\{D^T Pe\}\hat{d} - Hr - Te_1(t)] \tag{9.25}$$

such that $\dot{S}(t) = 0$. By substituting Equation 9.25 into Equation 9.24, the resulting equation is obtained as follows:

$$\dot{e}_1(t) = (A_m + BT)e_1(t). \tag{9.26}$$

If Equation 9.14 is satisfied, the dynamic response of $e_1(t)$ is globally and asymptotically stable, that is $\lim_{t\to\infty} e_1(t) = 0$. □

*Remark 9.2* Similar to Chapter 5 and Chapter 6, this chapter not only solves the dynamic tracking control problem through the designed control input, but also identifies the dynamic trajectory of the weight vectors related to the integral function (9.1) through the DNN identifier. The modified adaptive laws (6.18)-(6.20) can satisfy our demands better than those of [170, 171]. In Equation 6.20, $E[.]$ is applied because $W_2$ is a diagonal matrix.

*Remark 9.3* The design procedure can be described as follows:

1. For the target PDF, find the appropriate statistical information to characterize its main stochastic behaviors and determine the performance index.
2. For the measured output PDFs, use the two-step NN to approximate the performance function and identify the dynamical model relating the control input and weights.
3. Calculate the controller with variable structure form so as to achieve the tracking goal.

*Remark 9.4* From expression (9.17) for the input $u(t)$, the weights $W_1$, $W_2$, the states $x$, $\hat{x}$ and the functions $\phi(\hat{x})$, $\sigma(\hat{x}_\tau)$ can be proved bounded, so we can prove that the input is also bounded and that assumption 3 is reasonable. This is why we apply the projection arithmetics to the weights $W_1$, $W_2$. Based on Theorem 9.1, the states $x$, $\hat{x}$ can also be proved bounded. Of course, for on-line identification, it is a limitation and may cause the recurrent proof process. However, in much academic research into NN control or adaptive control, the recurrent proof process exists and is difficult to avoid. This problem will be considered carefully as a future research topic.

## 9.5 An Illustrative Example

In many practical processes, such as particle distribution control problems, the shapes of the measured output PDF have two or three peaks. Suppose that the output PDFs can be approximated using the B-spline models described by Equation 9.2 with $n = 3$, $y \in [0, 1.5]$ and for $i = 1, 2, 3$

$$c_i(y) = \begin{cases} |\sin 2\pi y| & y \in [0.5(i-1); 0.5i] \\ 0 & y \in [0.5(j-1); 0.5j] \quad i \neq j. \end{cases} \tag{9.27}$$

For simplicity, in the simulation it is assumed that the target PDF can be denoted by

$$\gamma_g(y) = \frac{1}{\sqrt{2\pi}b} e^{(-\frac{(y-a)^2}{2b^2})}$$

where $a = 0.5 + \ln\sqrt{2\pi} + \frac{4}{\pi}$, $b = 1$. From Equation 9.1, it can be obtained that

$$\Sigma_{i=1}^{3}(V_{gi}c_i(y)) = Q_1\gamma_g(y)ln(\gamma_g(y)) + Q_2 y\gamma_g(y)$$

where $Q_1 = 1$, $Q_2 = 1$. As a result, we have

$$\Sigma_{i=1}^{3}(V_{gi}\int_{-\infty}^{\infty}(c_i(y)\mathrm{d}y)) = (-\frac{1}{2})(1 + \ln(2\pi b^2)) + a. \tag{9.28}$$

Because $\int_{-\infty}^{\infty} c_i(y)\mathrm{d}y = \frac{1}{\pi}$ $i = 1,2,3$, it can be shown that the reference weights corresponding to the target statistical information satisfy the condition $\Sigma_{i=1}^{3}(V_{gi}) = 4$. As such, the reference weights can be denoted $V_g = [1\ \ 2\ \ 1]^T$.

In the simulation, it is assumed that the unknown dynamic model is given by

$$\dot{x}(t) = f(x) + g(x)u(t) + d_1, \tag{9.29}$$

$$f(x) = \begin{bmatrix} -2x_1 - x_2 \\ cosx_2 \\ x_2 - 3x_3 \end{bmatrix}, \ g(x) = \begin{bmatrix} 2\,0\,0 \\ 0\,2\,0 \\ 0\,0\,3 \end{bmatrix}, \ d_1 = \begin{bmatrix} 0.5 \\ 0.5 \\ 0.5 \end{bmatrix}, \ x(0) = \begin{bmatrix} 1 \\ 3 \\ 2 \end{bmatrix}.$$

By solving LMI (9.9) it can be obtained that

$$P = \begin{bmatrix} 1.8788 & 0 & 0 \\ 0 & 1.8788 & 0 \\ 0 & 0 & 1.8788 \end{bmatrix}, \ U = \begin{bmatrix} 1.7589 & 0 & -0.2644 \\ 0 & 1.7589 & 0 \\ 0.2644 & 0 & 1.7589 \end{bmatrix}.$$

For DNNs model (9.5), we select

$$\sigma(x_i) = \phi(x_i) = \frac{2}{1 + \mathrm{e}^{-0.5x_i}} + 0.5 \quad i = 1,2,3,$$

$$A = \begin{bmatrix} -2 & 0 & -2 \\ 0 & -2 & 0 \\ 2 & 0 & -2 \end{bmatrix}, \ B = \begin{bmatrix} 1\,0\,0 \\ 0\,1\,0 \\ 0\,0\,1 \end{bmatrix}, \ A_1 = \begin{bmatrix} -0.3 & 0 & 0 \\ 0 & -0.3 & 0 \\ 0 & 0 & -0.3 \end{bmatrix}, \ \hat{x}(0) = \begin{bmatrix} 1 \\ 3 \\ 2 \end{bmatrix}.$$

In adaptive laws (9.18)-(9.20), we select

$$M_1 = M_2 = 8, \ \ \varepsilon = 0.1, \ \ \gamma_i = 1 \ \ i = 1,2,3,$$

$$W_{10} = \mathrm{diag}[1,1,1], \ \ W_{20} = \mathrm{diag}[2,2,2].$$

In reference model (9.11), $A_m = \mathrm{diag}[-1,-1,-1], B_m = [1,2,1]^T, r = 1$. For sliding hyperplane $S(t)$, we select

$$G = \begin{bmatrix} 1\,0\,0 \\ 0\,1\,0 \\ 0\,0\,1 \end{bmatrix}, \quad T = \begin{bmatrix} -1 & -1 & 0 \\ 0 & 0 & -1 \\ -1 & 0 & -1 \end{bmatrix}.$$

With control input (9.17) and adaptive laws (9.18)-(9.20) applied, the states of the identified nonlinear system (9.29) are shown in Figure 9.1 and the state of DNNs (9.5) is shown in Figure 9.2. Figure 9.3 shows the states of the dynamic reference model. Figure 9.4 shows the sliding hyperplane $S(t)$. Finally, Figure 9.5 shows the 3D mesh plot of the integrated performance function. Figures 9.1-9.5 demonstrate that satisfactory tracking performance, identification capability and robustness are achieved.

On the other hand, in order to show the extensive applicability of the proposed method, we consider the following system:

$$\delta(\gamma, u(t)) = c_1(y)v_1 + c_2(y)v_1 + c_3(y)v_1$$

where $a$ and $b$ are set to -3 and +2, respectively, and

$$c_1(y) = [y^2 + 6y + 9]I_1 + [-y^2 - 3y - 1]I_2 + [y^2]I_3,$$

$$c_2(y) = [y^2 + 4y + 4]I_2 + [-y^2 - y + 1]I_3 + [y^2 - 2y + 1]I_4,$$

$$c_3(y) = [y^2 + 2y + 1]I_3 + [-y^2 + y + 1]I_4 + [y^2 - 4y + 4]I_5.$$

For this system, $I_i$ are the interval functions, defined as follows:

$$I_i(y) = \begin{cases} 1 & y \in [\lambda_i; \lambda_{i+1}) \\ 0 & elsewhere \end{cases}$$

where $\lambda_i = i - 4, \quad i = 1,2,3,4,5$. Other parameters are the same as for the previous design. Based on new B-spline models, the 3D mesh plot of the integrated performance function is showed in Figure 9.6. Figures 9.5-9.6 demonstrate that the satisfactory tracking performance can be obtained for different B-spline NN models.

## 9.6 Conclusions

Different from any other stochastic tracking problems, this chatper presents a new stochastic control setting where the tracking objective is some statistical information rather than a deterministic reference signal or a target PDF. A new two-step NN model including both static and dynamic parts is established to accomplish the STC problem. B-spline NNs are used to model the integrated function related to the tracked mean and entropy. DNNs describe the complex unknown nonlinear dynamics between the control input and the weight vectors related to the integrated function. An adaptive variable structure tracking controller based on the DNNs guaran-

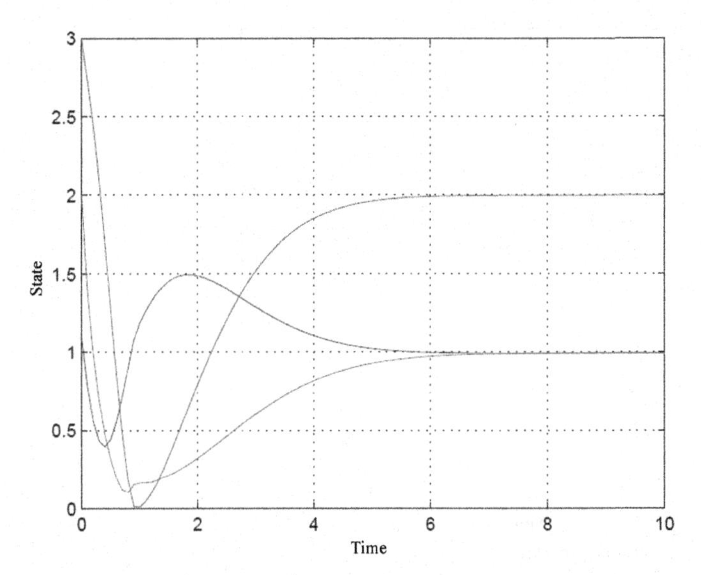


**Figure 9.1** States of the unknown nonlinear system

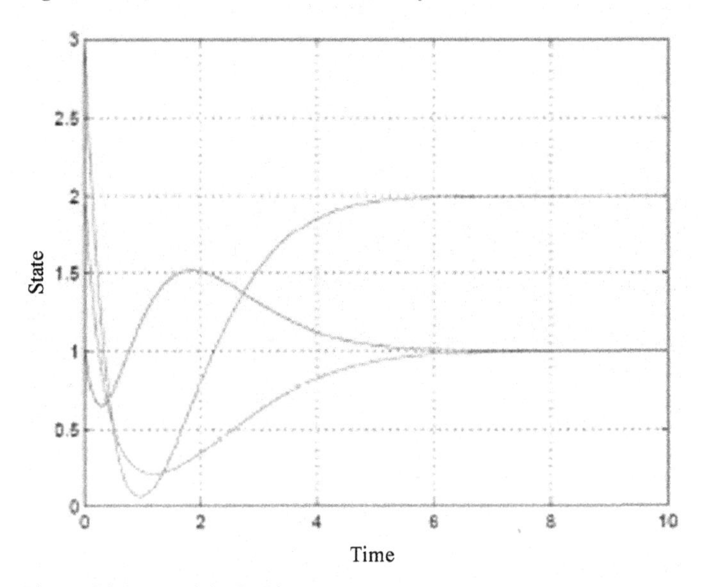


**Figure 9.2** States of the DNN

tees the tracking performance. Simulations are given to demonstrate the efficiency of the proposed approach.

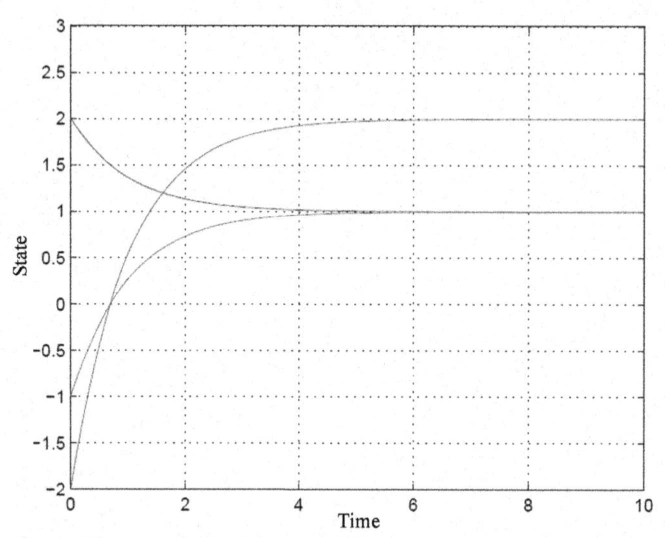


**Figure 9.3** States of the reference model

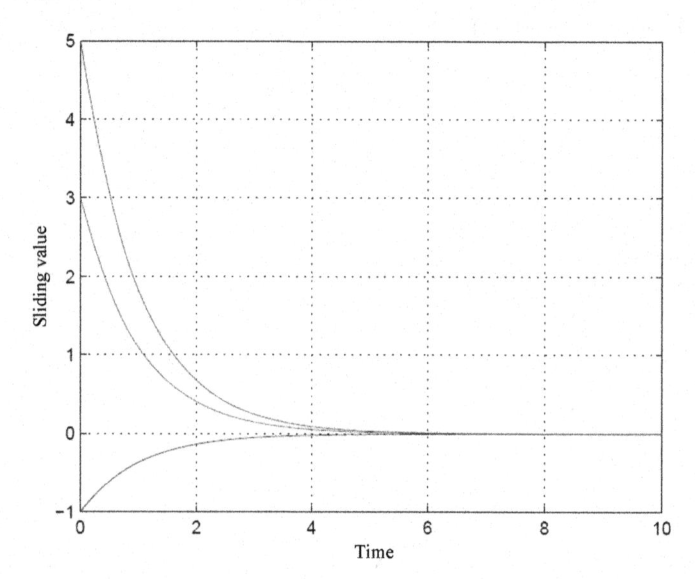


**Figure 9.4** Sliding function trajectory

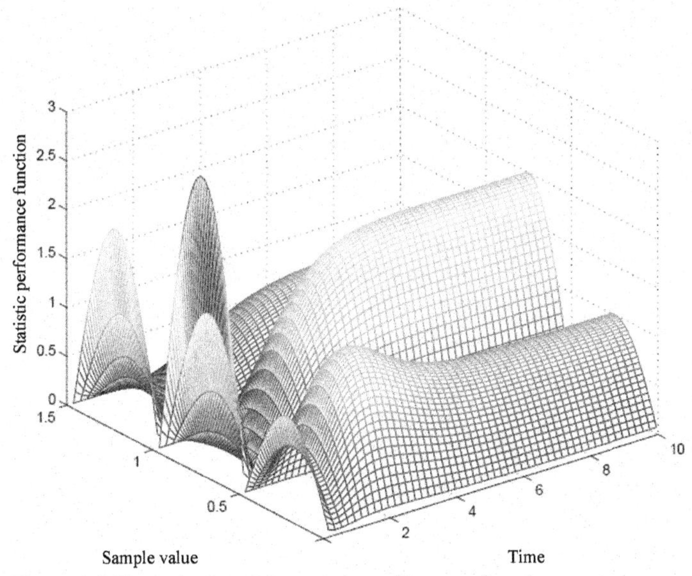


**Figure 9.5** 3D mesh plot of the statistic performance function

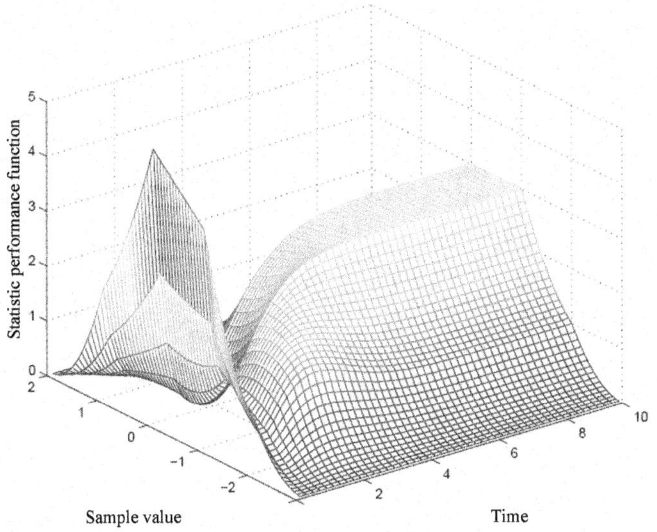


**Figure 9.6** 3D mesh plot of the statistic performance function

# Part IV
# Fault Detection and Diagnosis for Stochastic Distribution Control Systems

Safety and reliability are of paramount importance for practical processes. As a result, Fault detection and diagnosis (FDD) theory has been developed in the past three decades (see [6, 41, 48, 84–86, 121, 192, 197] and [201] for surveys). For stochastic systems, the approaches used thus far include the system identification technique [84] and statistical approaches based on the Bayesian theorem, likelihood methods, and hypothesis test techniques [6]. Also, filters or observers have been widely used to generate the residual signal for fault detection and estimation purposes (see [20, 48]). Generally, the observer-based or filter-based FDD methodologies have been developed along with the observer or filter design theory, and many of them have been applied to practical processes successfully.

Until now, most of the existing filter-based or observer-based FDD results for stochastic systems have been concerned only with Gaussian variables, where mean or variance was the objective for optimization. However, non-Gaussian variables exist widely in many complex stochastic systems due to the presence of nonlinearities, for which the classical filtering approaches are insufficient, especially for variables with asymmetric and multiple-peak stochastic distributions (see [62, 63, 76] and [157]). This is particularly true for stochastic distribution systems where the output is represented as the measured PDFs. Indeed, along with the continued and fast improvements of advanced instruments and data processing techniques, in practice the measurements for general filter design can be the stochastic distributions of the system output rather than its values. Typical examples include the particle-size distribution systems in chemical processing and the combustion flames distribution process [63, 157, 161]. As such, new filter- or observer-based FDD design algorithms are required for general stochastic systems using the output stochastic distributions.

For the SDC systems discussed in Part II, two main procedures were included [157]. The first is to use a B-spline expansion technique to model the measurable output PDFs, where PDFs can be represented by the weight dynamics corresponding to some basis functions. The second step is to establish a further dynamical model relating the input and dynamical weight vector of the B-spline expansion. In this part, square root B-spline expansions are adopted for the measured output PDFs of general stochastic systems, and nonlinear weight dynamical models are considered instead of linear ones. In Chapter 10, a new fault detection method using an augmented Lyapunov functional approach is presented. With the guaranteed cost performance index used as the objective function, an optimization algorithm with LMI constraint is applied to minimize the threshold value. This can improve residual signal sensitivity to the faults. Up to now few results have been seen focusing on FDD problems for the discrete-time SDC systems, where the nonlinearity and time delays are included and the measurement will be a nonlinear function of the state. To overcome the obstacles and improve FDD performances, in this part a discrete-time dynamical weighting system with nonlinearity, uncertainty and time delays is established, where nonlinearity exists in both the dynamical equation and the measurement one. It is noted that even in the context of conventional FDD problems, few available results can be used for such a complex model. In this part, robust guaranteed cost performance is introduced for the weighting model to enhance FDD performance. Feasible approaches are given to design the FDD algorithms.

# Chapter 10
# Optimal Continuous-time Fault Detection Filtering

## 10.1 Introduction

The purpose of fault detection is to determine the occurrence of faults in the system. For dynamic stochastic systems, the filter-based fault detection approach has been shown to be effective when the variables are assumed to be Gaussian [20, 110]. However, in many practical processes, non-Gaussian variables exist in many stochastic systems due to nonlinearity, and these may possess asymmetric and multiple-peak stochastic distributions [93].

In output PDFs shape control, B-spline NN have been used for modeling the output PDFs in [157]. The motivation for fault detection via the output PDFs from the retention system in paper making was first studied in [161], where the weight dynamical system was assumed to be a precise linear model. Linear mappings cannot change the shape of output PDFs, which implies that the fault cannot be detected through shape change of the PDFs. To meet the requirement in complex processes, a kind of observer-based fault detection algorithm has been established in [66] and [73]. It is noted that the research framework of stochastic distribution systems (SDSs) was established, and the nonlinear weighting system was considered in [66] and [73]. However, when the fault is small, the fault cannot be detected using the algorithms in [66] and [73]. This is because only the uniform boundedness of the estimation error could be guaranteed in [66], which led to some conservative criteria. It is difficult to compute the optimal threshold using the FDD filtering given in [73].

This chapter presents a new fault detection method using an augmented Lyapunov functional approach. A two-step NN modeling approach has been established. A static NN based square root B-spline approximation technique is applied to model the PDFs and a DNN model is used to describe the relationship between the input and weights. With the guaranteed cost performance index used as the objective function, an optimization algorithm with LMI constraint is applied to minimize the threshold value. This can improve residual signal sensitivity to the faults.

## 10.2 Problem Formulation

First, we briefly review the static NN model for PDFs. The square-root B-spline approximation technique presented in [73] is used to formulate the output PDFs with the dynamic weight. This is essential in solving the fault detection problem for SDCs.

For a dynamic stochastic system, $y(t) \in [a,\ b]$ is the output of the dynamic stochastic system, the conditional probability of output $y(t)$ lying inside $[a,z]$ is defined by $P\{a \le y(t) \le z\} = \int_a^z \gamma(\zeta, u(t), F)\mathrm{d}\zeta$, where $z$ is the variable defined on $[a,\ b]$, $\gamma(z,u(t),F)$ is the output PDFs, $u(t) \in R^m$ is control input and $F$ is the fault vector to be detected. For example, in paper making systems, $F$ can be regarded as caused by the change in strength of the fluid turbulence in the headbox approaching system (see [161]). In [63] and [73], some B-spline approximation models have been used to approximate $\gamma(z,u(t),F)$. In this chapter, we use the following square root B-spline approximation technique model:

$$\sqrt{\gamma(z,u(t),F)} = \sum_{i=1}^{n} v_i(u,F)b_i(z) \tag{10.1}$$

where $b_i(z)(i = 1,2,...,n)$ are pre-specified basis functions defined on $[a,b]$, and $v_i(u(t),F)(i = 1,2,...,n)$ are the corresponding weights of such an expansion. Denote

$$B_0(z) = [b_1(z)\ \ b_2(z) \cdots\ \ b_{n-1}(z)]^T,$$
$$V(t) = V(u(t),F) = [v_1\ \ v_2 \cdots\ \ v_{n-1}]^T$$

and let

$$\Lambda_1 = \int_a^b B_0(z)B_0^T(z)\mathrm{d}z,\ \Lambda_2 = \int_a^b B_0^T(z)b_n(z)\mathrm{d}z,$$

$$\Lambda_3 = \int_a^b b_n^2(z)\mathrm{d}z \neq 0,\ \Lambda_0 = \Lambda_1\Lambda_3 - \Lambda_2\Lambda_2^T.$$

Furthermore, it can be verified that Equation 10.1 can be rewritten as (see [66] for details)

$$\sqrt{\gamma(z,u(t),F)} = B^T(z)V(t) + h(V(t))b_n(z) \tag{10.2}$$

where

$$B^T(z) = B_0^T(z) - \frac{\Lambda_2}{\Lambda_3}b_n(z),$$

$$h(V(t)) = \frac{\sqrt{\Lambda_3 - V^T(t)\Lambda_0 V^T(t)}}{\Lambda_3}. \tag{10.3}$$

For $h(V(t))$ denoted by Equation 10.3, it is assumed that the Lipschitz condition can be satisfied within its operating region, i.e., for any $V_1(t)$ and $V_2(t)$, there exists a known matrix $U_1$ such that

$$\| h(V_1(t)) - h(V_2(t)) \| \leq \| U_1(V_1(t) - V_2(t)) \| \tag{10.4}$$

where $\| \cdot \|$ denotes the Euclidean norm.

Second, we will find the relationship between the input and the weights related to the PDFs, which corresponds to a further modeling procedure. Most published results concern only linear precise models, while practically the relationships from control input $u(t)$ to weight vector $V(t)$ should be nonlinear dynamics and subject to some uncertainties [63]. In this chapter, we can adopt NN modeling for the the second-step modeling procedure. As such, the following nonlinear dynamic model will be considered in this chapter:

$$\begin{cases} \dot{x}(t) = Ax(t) + Gg(x(t)) + Hu(t) + JF(t) \\ V(t) = Ex(t) \end{cases} \tag{10.5}$$

where $x(t) \in R^m$ is the unmeasured state, and $F(t)$ is the fault to be detected. $A$, $G$, $H$, $J$ and $E$ represent the known parametric matrices of the dynamic part of the weight system. The initial condition is defined by $x_0$. Similarly to [66] and [73], $g(x(t))$ is assumed to satisfy $g(0) = 0$ and the following norm condition

$$\| g(x_1(t)) - g(x_2(t)) \| \leq \| U_2(x_1(t) - x_2(t)) \| \tag{10.6}$$

for any $x_1(t)$ and $x_2(t)$, where $U_2$ is a known matrix. Inequalities (10.4) and (10.6) can be guaranteed by the property of functions $h(V(t))$ and $g(x(t))$ and the boundedness of $V(t)$. With Equation 10.5, output Equation 10.2 can be further rewritten as

$$\sqrt{\gamma(z,u(t),F)} = B^T(z)Ex(t) + h(Ex(t))b_n(z). \tag{10.7}$$

*Remark 10.1* Inequalities (10.4) and (10.6) are typically required in the literature on fault detection for nonlinear systems (see [35]). The assumptions will be used to simplify the observer design algorithm in the following sections.

## 10.3 Main Result

Generally speaking, a fault detection system consists of two parts: a residual generator and a residual evaluator including a threshold and a decision logic unit. For the purpose of residual generation, we construct the following nonlinear observer:

$$\dot{\hat{x}}(t) = A\hat{x}(t) + Gg(\hat{x}(t)) + Hu(t) + L\xi(t) \tag{10.8}$$

where the residual signal is defined in terms of output PDFs as

$$\xi(t) = \int_a^b \sigma(z) \left( \sqrt{\gamma(z,u(t),F)} - \sqrt{\hat{\gamma}(z,u(t))} \right) \mathrm{d}z$$

and

$$\sqrt{\hat{\gamma}(z,u(t))} = B^T(z)E\hat{x}(t) + h(E\hat{x}(t))b_n(z)$$

where $\hat{x}(t)$ is the estimated state, $L \in R^{m\times p}$ is the gain to be determined and $\sigma(z) \in R^{p\times 1}$ can be regarded as a pre-specified weighting vector defined on $[a,b]$. Different from most classical observer design methods [19], residual $\xi(t)$ is formulated as an integral with respect to the difference of the measured PDFs and the estimated PDFs.

By defining $e(t) = x(t) - \hat{x}(t)$, the estimation error system can be described by

$$\begin{aligned}\dot{e}(t) = {} & (A - L\Gamma_1)e(t) + G[g(x(t)) - g(\hat{x}(t))] \\ & -L\Gamma_2[h(Ex(t)) - h(E\hat{x}(t))] + JF(t)\end{aligned} \tag{10.9}$$

where

$$\Gamma_1 = \int_a^b \sigma(z)B^T(z)\mathrm{E}dz,\ \Gamma_2 = \int_a^b \sigma(z)b_n(z)dz. \tag{10.10}$$

It can be seen that

$$\xi(t) = \Gamma_1 e(t) + \Gamma_2(h(Ex(t)) - h(E\hat{x}(t))). \tag{10.11}$$

In order to detect the fault more sensitively, optimization techniques for the error systems are preferable. For this purpose, the following reference vector is constructed:

$$z(t) = C\xi(t) = C\Gamma_1 e(t) + C\Gamma_2(h(Ex(t)) - h(E\hat{x}(t))) \tag{10.12}$$

where $C$ is a selected weighting matrix. It is noted that the nonlinear function $z(t)$ includes only the measurable information related to the output PDFs. At this stage, we consider the sub-optimal guaranteed cost for the $H_2$ performance index as $J = ||z(t)||_2^2 = \int_0^\infty z^T(t)z(t)dt$ subject to the error system (10.9). Setting $\hat{x}(0) = 0$, hence $e(0) = x_0$. Thus, the problem of designing observer-based fault detection, which is one of the main objectives of this work, can be described as designing matrix $L$ such that

1. the error system (10.9) is asymptotically stable;
2. the $H_2$ performance index $J$ satisfies $J \leq J_b$ and $J_b$ is as small as possible.

The following result provides a design criterion to detect the fault using the detection observer (10.8) through the output PDFs.

**Theorem 10.1** For the parameters $\lambda_i > 0 (i = 1,2)$ and $\varepsilon$, if there exist matrices $P_1 > 0$, $P_2$, $R$ satisfying

$$\Pi = \begin{bmatrix} \Pi_{11} & \Pi_{12} & P_2^T G & -R\Gamma_2 & \Gamma_1^T C^T \\ * & \Pi_{22} & \varepsilon P_2^T G & -\varepsilon R\Gamma_2 & 0 \\ * & * & -\frac{1}{\lambda_2^2} I & 0 & 0 \\ * & * & * & -\frac{1}{\lambda_1^2} I & \Gamma_2^T C^T \\ * & * & * & * & -I \end{bmatrix} < 0 \tag{10.13}$$

where

$$\Pi_{11} = P_2^T A - R\Gamma_1 + A^T P_2 - \Gamma_1^T R^T + \frac{1}{\lambda_1^2} E^T U_1^T U_1 E + \frac{1}{\lambda_2^2} U_2^T U_2, \tag{10.14}$$

$$\Pi_{12} = \varepsilon A^T P_2 - \varepsilon \Gamma_1^T R^T + P_1 - P_2^T, \tag{10.15}$$

$$\Pi_{22} = -\varepsilon P_2^T - \varepsilon P_2, \tag{10.16}$$

then in the absence of $F$, the error system (10.9) with gain $L = P_2^{-T} R$ is stable and satisfies

$$J \leq x_0^T P_1 x_0. \tag{10.17}$$

*Proof.* Define $\tilde{g} := g(x(s)) - g(\hat{x}(s))$, $\tilde{h} := h(Ex(s)) - h(E\hat{x}(s))$ and denote the Lyapunov function candidate as follows:

$$\Phi(t) = \bar{e}^T(t) SP\bar{e}(t) + \frac{1}{\lambda_1^2}\int_0^t [\| U_1 Ee(s) \|^2 - \| \tilde{h} \|^2] ds + \frac{1}{\lambda_2^2}\int_0^t [\| U_2 e(s) \|^2 - \| \tilde{g} \|^2] ds \tag{10.18}$$

where

$$S = \begin{bmatrix} I & 0 \\ 0 & 0 \end{bmatrix}, \; P = \begin{bmatrix} P_1 & 0 \\ P_2 & \varepsilon P_2 \end{bmatrix}, \; \bar{e}(t) = \begin{bmatrix} e(t) \\ \dot{e}(t) \end{bmatrix},$$

$$P_1 > 0, \; SP = P^T S^T \geq 0.$$

Following Equations 10.4 and 10.6 yields $\Phi(t) \geq 0$ for all arguments. It is noted that $\bar{e}^T(t) SP\bar{e}(t)$ is actually $e^T(t) P_1 e(t)$. Hence, differentiating $\bar{e}^T(t) SP\bar{e}(t)$ with respect to gives

$$\frac{d}{dt}\{\bar{e}^T(t) SP\bar{e}(t)\} = 2e^T(t) P_1 \dot{e}(t) = 2\bar{e}^T(t) P^T \begin{bmatrix} \dot{e}(t) \\ 0 \end{bmatrix}.$$

On the other hand, in the absence of $F(t)$, the state Equation 10.9 ensures that

$$\alpha(t) := -\dot{e}(t) + (A - L\Gamma_1)e(t) + G\tilde{g} - L\Gamma_2 \tilde{h} = 0.$$

Then along the trajectory of Equation 10.9 in the absence of $F$, it can be shown that

$$
\begin{aligned}
\dot{\Phi}(t) &= 2\bar{e}^T(t)P^T\begin{bmatrix}\dot{e}(t)\\ \alpha(t)\end{bmatrix} + \frac{1}{\lambda_1^2}[\| U_1Ee(t)\|^2 - \|\tilde{h}\|^2] + \frac{1}{\lambda_2^2}[\| U_2e(t)\|^2 - \|\tilde{g}\|^2]\\
&\leq 2e^T(t)P_1\dot{e}(t) - 2e^T(t)P_2^T\dot{e}(t) + 2e^T(t)(P_2^TA - R\Gamma_1)e(t) + 2e^T(t)P_2^TG\tilde{g}\\
&\quad -2e^T(t)R\Gamma_2\tilde{h} - 2\varepsilon\dot{e}^T(t)P_2^T\dot{e}(t) + 2\varepsilon\dot{e}^T(t)(P_2^TA - R\Gamma_1)e(t) + 2\varepsilon\dot{e}^T(t)P_2^TG\tilde{g}\\
&\quad -2\varepsilon\dot{e}^T(t)R\Gamma_2\tilde{h} - \frac{1}{\lambda_1^2}\tilde{h}^T\tilde{h} - \frac{1}{\lambda_2^2}\tilde{g}^T\tilde{g} + e^T(t)[\frac{1}{\lambda_2^2}U_2^TU_2 + \frac{1}{\lambda_1^2}E^TU_1^TU_1E]e(t)\\
&= \eta^T(t)\Pi_0\eta(t)
\end{aligned} \tag{10.19}
$$

where $R = P_2^TL,\ \eta(t) = [e^T(t)\ \ \dot{e}^T(t)\ \ \tilde{g}^T\ \ \tilde{h}^T]^T$ and

$$
\Pi_0 = \begin{bmatrix}\Pi_{11} & \Pi_{12} & P_2^TG & -R\Gamma_2\\ * & \Pi_{22} & \varepsilon P_2^TG & -\varepsilon R\Gamma_2\\ * & * & -\frac{1}{\lambda_2^2}I & 0\\ * & * & * & -\frac{1}{\lambda_1^2}I\end{bmatrix}.
$$

$\Pi_{11}$, $\Pi_{12}$, $\Pi_{22}$ are defined in Equations 10.14-10.16. Therefore, it can be claimed that $\dot{\Phi}(t)$ always holds and the estimation error system is stable. Furthermore, it is noted that

$$
\begin{aligned}
z^T(t)z(t) &= e^T(t)\Gamma_1^TC^TC\Gamma_1e(t) + 2e^T(t)\Gamma_1^TC^TC\Gamma_2\tilde{h}\\
&\quad +\tilde{h}^T\Gamma_2^TC^TC\Gamma_2\tilde{h}.
\end{aligned} \tag{10.20}
$$

From Equations 10.19 and 10.20, we have

$$
z^T(\tau)z(\tau) + \dot{\Phi}(\tau) \leq \eta^T(\tau)(\Pi_0 + \Pi_1)\eta(\tau)
$$

where

$$
\Pi_1 = \begin{bmatrix}\Gamma_1^TC^TC\Gamma_1 & 0 & 0 & \Gamma_1^TC^TC\Gamma_2\\ * & 0 & 0 & 0\\ * & * & 0 & 0\\ * & * & * & \Gamma_2^TC^TC\Gamma_2\end{bmatrix}.
$$

According to Equation 10.13, one obtains

$$
z^T(\tau)z(\tau) + \dot{\Phi}(\tau) \leq 0.
$$

By integrating both sides of the above inequality from 0 to $T_f$, we obtain

$$
\int_0^{t_f} z^T(t)z(t)\mathrm{d}t \leq \Phi(0) - \Phi(T_f).
$$

When $T_f \to \infty$, we get

$$
||z(t)||_2^2 \leq \Phi(0)
$$

where $\Phi(0) = x_0^T P_1 x_0$ which leads to Equation 10.17 by using the definition of $J$. This completes the proof. □

If we choose $\Phi(t) = e^T(t)Pe(t) + \frac{1}{\lambda_1^2}\int_0^t[\| U_1 Ee(s) \|^2 - \| \tilde{h} \|^2]ds + \frac{1}{\lambda_2^2}\int_0^t[\| U_2 e(s) \|^2 - \| \tilde{g} \|^2]ds$ as [66], where $P > 0$, the following result can be obtained by similar proof procedure of Theorem 10.1.

**Corollary 10.1** For the parameters $\lambda_i > 0 (i = 1,2)$, if there exist matrices $P > 0$, $R$ satisfying

$$\begin{bmatrix} \hat{\Pi}_{11} + \frac{1}{\lambda_2^2}U_2^T U_2 & P_2^T G & -R\Gamma_2 & \Gamma_1^T C^T \\ * & -\frac{1}{\lambda_2^2}I & 0 & 0 \\ * & * & -\frac{1}{\lambda_1^2}I & \Gamma_2^T C^T \\ * & * & * & -I \end{bmatrix} < 0 \tag{10.21}$$

where

$$\hat{\Pi}_{11} = PA - R\Gamma_1 + A^T P - \Gamma_1^T R^T + \frac{1}{\lambda_1^2}E^T U_1^T U_1 E,$$

then in the absence of $F$, the error system (10.9) with gain $L = P^{-1}R$ is stable and satisfies $J \leq x_0^T P x_0$.

*Remark 10.2* Recently, the fault detection problem was considered using the PDF in [66] and [73]. It is worth pointing out that only the uniform boundedness of the estimation error $e(t)$ could be guaranteed in [66]. However, Theorem 10.1 can guarantee the cost $J$ is bounded, which includes the boundedness of $e(t)$ as a special case. And, it is difficult to obtain the optimal threshold value using the algorithms in [73].

*Remark 10.3* In the derivation of Theorem 10.1, a new Lyapunov function (10.18) is introduced, which involves the tuning parameter $\varepsilon$ and slack variable $P_2$. The advantage of this function is that it separates Lyapunov function matrix $P_1$ from $A$ and $L$, which make the solution of inequality (10.13) more flexible. Moreover, it is pointed out that inequality (10.21) can be covered by setting $\varepsilon = 0$, $P_2 = P_1 = P$ in Equation 10.13.

After designing the fault detection observer, the next important task for fault detection is the evaluation of the generated residual, including a threshold and a decision logic unit. In this case, we choose $||z(t)||$ and $\beta = \sqrt{x_0^T P_1 x_0}$ as residual evaluation function and threshold, respectively, and, based on this, use the following logical relationship for fault detection:

$||z(t)|| > \beta \Rightarrow$ faults $\Rightarrow$ alarm,
$||z(t)|| \leq \beta \Rightarrow$ no faults.

From the above logical relationship, it is clear that the fault detection performance may be improved by minimizing the threshold $\beta$. The following theorem is applied to optimal $x_0^T P_1 x_0$.

**Theorem 10.2** Given the parameters $\lambda_i$ $(i = 1,2)$*and* $\varepsilon$, if the following optimization problem:

$$\min_{P_1,\ P_2,\ R} \alpha \tag{10.22}$$

subject to

$$(i)\ LMI\ (13)\quad (ii)\ \begin{bmatrix} -\alpha & x_0^T P_1 \\ P_1 x_0 & -P_1 \end{bmatrix} < 0 \tag{10.23}$$

has solutions with positive definite matrix $P_1$, matrices $P_2$, $R$ and positive scalar $\alpha$, then $L = P_2^{-T} R$ is an optimal guaranteed cost gain which ensures minimization of the guaranteed cost $x_0^T P_1 x_0$ for error system (10.9).

The proof of Theorem 10.2 is omitted for brevity. Corresponding to Corollary 10.1, the following corollary is also provided.

**Corollary 10.2** Given the parameters $\lambda_i$ $(i = 1,2)$, if the following optimization problem:

$$\min_{P,\ R} \alpha$$

subject to

$$(i)\ LMI\ (21)\quad (ii)\ \begin{bmatrix} -\alpha & x_0^T P \\ P x_0 & -P \end{bmatrix} < 0$$

has solutions with matrices $P > 0$, $R$ and a constant $\alpha > 0$, then $L = P^{-1} R$ is an optimal guaranteed cost gain which ensures minimization of the guaranteed cost $x_0^T P x_0$.

*Remark 10.4* The inequality (10.13) is an LMI with tuning parameter. By tuning the free parameter $\varepsilon$ and combining with the optimization algorithm in Theorem 10.2, we can provide a lower bound of the threshold $\beta$ and more feasible observer design than Corollary 10.2.

## 10.4 Numerical Examples

In this section an example is provided to illustrate the usefulness of our method. Consider the particle distribution control problems given in [63]. Particulate processes are often modeled using the following form of the population balance equation, which describes the evolution of particle distribution:

$$\frac{\mathrm{d}W_j(t)}{\mathrm{d}t} = H[W_j(t), z(t), u(t), F],\ j \in [1,\ n],$$

$$\frac{dz(t)}{dt} = G[W_j(t), z(t), u(t), F]$$

where $W_j(t)$ is the value of the particle distribution $W(\xi, t)$ at the discretization point $\xi_j$, $H(.)$ and $G(.)$ are nonlinear algebraic functions, which depend on the differential variables $z(t)$, input $u(t)$ and the fault $F$. By defining $x^T = [W_1,\ W_2, ...,\ W_n, z^T]$, part Jacobian linearization of this model at a steady-state operating point leads to a nonlinear state-space form Equation 10.5. The shapes of measured output PDF for Equation 10.5 usually have two or three peaks. Suppose these output PDFs can be approximated using a square root B-spline model $\sqrt{\gamma(z, u(t))} = \sum_{i=1}^{3} v_i(u(t), F) b_i(z)$, where $z$ is defined in $[0, 1.5]$ and

$$b_i = \begin{cases} |\sin 2\pi z|, & z \in [0.5(i-1),\ 0.5i] \\ 0, & others \end{cases}$$

for $i = 1,\ 2,\ 3$. Furthermore, it can be seen that

$$\Lambda_1 = \begin{bmatrix} 0.25 & 0 \\ 0 & 0.25 \end{bmatrix},\ \Lambda_2 = \begin{bmatrix} 0 & 0 \end{bmatrix},\ \Lambda_3 = 0.25.$$

It is assumed that the identified weighting system is formulated by Equation 10.5 with the following coefficient matrices:

$$A = \begin{bmatrix} -0.5 & 0.3 \\ 0 & -1.3 \end{bmatrix},\ G = \begin{bmatrix} 0 & 0 \\ 0 & 0.1 \end{bmatrix},\ E = \begin{bmatrix} 1 & 0 \\ 0 & 1 \end{bmatrix},$$

$$H = \begin{bmatrix} 0.2 & 0 \\ 0 & -0.3 \end{bmatrix},\ J = \begin{bmatrix} 0.9 \\ 0.9 \end{bmatrix}.$$

The upper bounds of nonlinearity are denoted by $U_1 = \mathrm{diag}\{1,\ 1\}$, $U_2 = \mathrm{diag}\{0,\ 0.5\}$. For the reference output, $C = 1$. It can be tested that $\Gamma_1 = [\frac{1}{\pi}\ \ \frac{1}{\pi}]$, $\Gamma_2 = \frac{1}{\pi}$ for $\sigma(z) = 1$. In the simulation, the initial condition of the system state and its estimation are selected as $x(t) = [0.2\ \ -0.1]^T, \hat{x}(0) = [0\ \ 0]$ with the parameters being given as $\lambda_1 = 1,\ \lambda_2 = 1,\ \varepsilon = 0.1$. The fault is assumed given by

$$F(t) = \begin{cases} 0, & t < 10, \\ 1.6, & 10 \le t \le 20, \\ 0, & t \ge 20. \end{cases}$$

The same threshold can be calculated ($\beta = 0.0332$) in Theorem 10.2 and Corollary 10.2, but observer gain $L = [26.9982\ \ 10.8077]$ and $L = 10^5 \times [2.6900\ \ 1.4396]$ are obtained by Theorem 10.2 and Corollary 10.2, respectively. It is clear that the method of Theorem 10.2 is more feasible than that of Corollary 10.2.

In Figure 10.1, the 3D mesh plot shows changes in the measured output PDFs. Figure 10.2 shows the residual signal when the fault occurs from 10 to 20 s. Figure

10.3 shows the evolution of the residual evaluation function and the threshold. From Figure 10.3, it can be seen that the fault is detected 2 s after its occurrence.

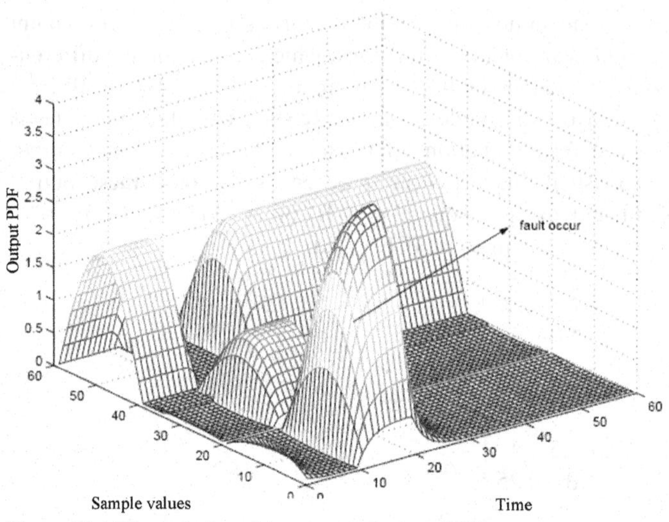


**Figure 10.1** 3D mesh plot of the measured output PDFs

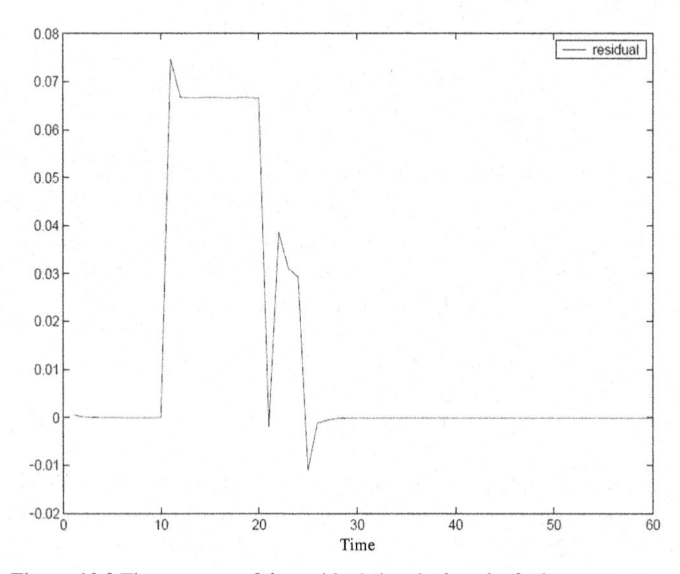


**Figure 10.2** The response of the residual signal when the fault occurs

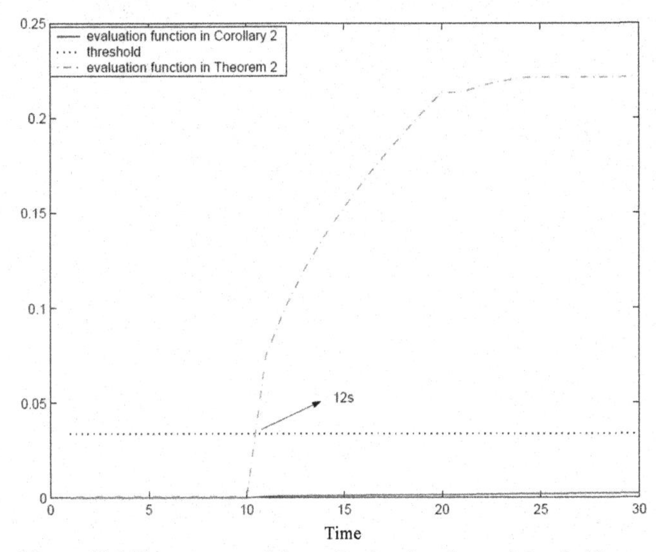


**Figure 10.3** The response of the evaluation function and threshold

## 10.5 Conclusions

In this chapter the fault detection problem is investigated for stochastic distribution systems using two-step NN modeling procedures. Based on LMI techniques, a new criterion for the existence of a reliable guaranteed cost observer is obtained for fault detection. By introducing a tuning parameter, the detection sensitivity performance is improved by minimizing the guaranteed cost. A simulation is given to demonstrate the efficiency of the proposed approach.

To ensure simplicity of the formulation, in this chapter robustness issues have not been discussed. When there exist model uncertainties, the gain matrix $L$ should be designed so that the residual signal $\xi(t)$ is sensitive to the fault but insensitive to the model uncertainties. This needs future investigation.

# Chapter 11
# Optimal Discrete-time Fault Detection and Diagnosis Filtering

## 11.1 Introduction

For the FDD problem of stochastic plants, one main approach is the filter-based or observer-based methodology, where the (extended) Kalman filtering results were widely used and Gaussian systems were considered (see [20, 41]). However, nonlinearity may lead to non-Gaussian outputs, where (especially for asymmetric distributions with multiple peaks) mean and variance are insufficient to characterize their statistical behavior precisely (see [63, 66, 157, 161]). It is noted that many effective robust and nonlinear filtering and filter-based FDD approaches have been provided (see [55, 110, 144, 176, 195, 203] and references therein) to date. For example, in [55], robust performance optimization has been applied to the filtering problems of linear systems. In [176], feasible filtering approaches based on algebraic algorithms have been provided for linear plants with uncertainties and time delays. In [110, 144], it has been shown how robust filtering and robust control approaches can be used for FDD problems. However, few results can be found to solve FDD problems for the transformed discrete-time system, when nonlinearity and time delays are included and measurements are a nonlinear function of the state.

To overcome the obstacles and improve FDD performance, a discrete-time dynamical weighting system with nonlinearity, uncertainty and time delays is established. As such, the problem for dynamic stochastic systems can be transformed into a classical discrete-time nonlinear FDD problem, where nonlinearity exists in both the dynamical equation and the measurement one. In this chapter, robust guaranteed cost performance is introduced for the weighting model to enhance FDD performance. Feasible approaches are given to design the FDD filter by means of optimization techniques in terms of LMIs, with which the stochastic faults can be detected and diagnosed using the measured output PDFs.

## 11.2 Formulation of the FDD Problem with Guaranteed Cost Performance

The square root B-spline expansions have been effectively used in PDF control and the corresponding FDD problems (see [63, 66]). However, most of them only consider continuous-time systems while, practically, discrete-time models are usually more useful in both the modeling and the actuation processes. In this section, different from the recent results in [63, 66], we introduce a discrete-time B-spline square root model to approximate the output PDFs and then formulate a discrete-time nonlinear model for the weighting vectors.

Consider a discrete-time dynamic stochastic system where $u(k) \in R^m$ is the input, $y(k) \in [a,b]$ is the output, and $F(k)$ is the fault. The objective in the FDD context is to use $\gamma(z,u(k),F(k))$ to design the filter gain such that $F(k)$ can be detected and diagnosed. For the considered filter-based FDD problem, $u(k)$ is irrelevant and will be neglected in the following arguments and $\gamma(z,u(k),F(k))$ will be abbreviated by $\gamma(z,F,k)$.

Denote $b_i(z)(i=1,2,\cdots n)$ as a set of pre-specified basis functions on $[a,b]$, with $v_i(F,k)(i=1,2,\cdots n)$ the corresponding weights. In this chapter, the square root B-spline model is adopted to enhance the robustness of the model (see [63, 66] for the corresponding continuous-time case). At sample time $k$, the square root of the conditional PDFs can be approximated as

$$\sqrt{\gamma(z,F,k)} = B_0(z)V(k) + h_0(V(k))b_n(z) \tag{11.1}$$

where

$$\begin{aligned} B_0(z) &= [b_1(z) \ \ b_2(z) \ \ \cdots \ \ b_{n-1}(z)], \\ V(k) &= [v_1(F,k) \ \ v_2(F,k) \ \ \cdots \ \ v_{n-1}(F,k)]^T, \\ h_0(V(k)) &= \frac{\sqrt{\Lambda_3 - V^T(k)\Lambda_0 V(k)} - \Lambda_2 V(k)}{\Lambda_3}, \end{aligned} \tag{11.2}$$

and

$$\Lambda_1 = \int_a^b B_0^T(z)B_0(z)\mathrm{d}z, \ \ \Lambda_2 = \int_a^b B_0^T(z)b_n(z)\mathrm{d}z, \ \ \Lambda_3 = \int_a^b b_n^2(z)\mathrm{d}z. \tag{11.3}$$

Different from [63, 66, 161], we adopt the following model:

$$\sqrt{\gamma(z,F,k)} = B(z)V(k) + h(V(k))b_n(z)$$

instead of Equation 11.1, where

$$B(z) = B_0(z) - \frac{\Lambda_2}{\Lambda_3}b_n(z), \ \ h(V(k)) = \frac{\Lambda_3 - V^T(k)\Lambda_0 V(k)}{\Lambda_3}. \tag{11.4}$$

For $h(V(k))$ denoted by Equation 11.3, it is assumed that the Lipschitz condition is satisfied within its operating region, which means that for any $V_1(k)$ and $V_2(k)$, there exists a known matrix $U_1$ such that

$$\|h(V_1(k)) - h(V_2(k))\| \leq \|U_1(V_1(k) - V_2(k))\|. \tag{11.5}$$

After B-spline expansions to the output PDFs are made, the dynamics between the output PDFs and the input can be further expressed as the relationship between the input and the weights. The procedure corresponds to an identification method that has also been used for PDF control and entropy optimization (see [63, 157]), also FDD problems (see [66, 161]). To meet the practical requirements, in this chapter we consider a discrete-time dynamic model with time delays and nonlinearity as follows:

$$\begin{cases} x(k+1) = Ax(k) + A_d x(k-d) + Gg(x(k)) + Hu(k) + DF(k) \\ V(k) = Ex(k) \end{cases} \tag{11.6}$$

where $x(k) \in R^m$ is the unmeasured state, $V(k)$ is denoted in Equation 11.2 as the weights, $F(k)$ is the fault to be detected, and $d$ is the known time delay. $A, A_d, G, H, D$ and $E$ represent the known parametric matrices. $g(x(k))$ is a known nonlinear function which satisfies the following norm condition:

$$\|g(x_1(k)) - g(x_2(k))\| \leq \|U_2(x_1(k) - x_2(k))\| \tag{11.7}$$

for any $x_1(k)$ and $x_2(k)$, where $U_2$ is a known matrix. The initial condition is denoted by $x(k) = \varphi(k), k = -d, -d+1, \cdots, -1, 0$. In fact, due to the measured error for the weights, one can suppose that the initial value belongs to the set $S = \{\varphi(-j) : \varphi(-j) = S_j \eta_j, \eta_j^T \eta_j \leq 1, j = 0, 1, \cdots, d\}$, where $S_j$ represents the known matrix and $\eta_j$ is the uncertain but bounded vector.

It is shown that model (11.6) can be considered as an uncertain nonlinear time delayed discrete-time system with non-zero initial conditions. With model (11.6), output Equation 11.1 can be rewritten as a nonlinear function of $x(k)$ as follows:

$$\sqrt{\gamma(z, F, k)} = B(z)Ex(k) + h(Ex(k))B_n(z). \tag{11.8}$$

*Remark 11.1* Compared with the models considered in [63, 66, 157, 161], there are several features of the proposed formulation. First, the proposed discrete-time square root B-spline model is more practically reasonable and better suited to computer control. In the models adopted in [63, 66], nonlinear function $h_0(V(k))$ was used instead of $h(V(k))$, which can lead to conservative results. Second, to meet the requirement of practical modeling and applications, time delays have been included in the discrete-time weighting systems.

In order to decrease the conservativeness of the proposed FDD approaches, in the following we will establish the guaranteed cost FDD filtering design methods. It is noted that even for conventional FDD problems on discrete-time systems with time delays, fewer available methodologies have been provided for nonlinear systems with time delays (see [41, 66, 144, 176, 195]).

## 11.3 The Optimal Fault Detection Filter Design

Different from most classical observer or filter design methods (see [55, 144, 195]), the information for filter design is the distance between the measured PDFs and the estimated PDFs from the filter (see [66]). In order to detect the fault based on the changes of output distributions, we construct the following vector as the measurement output and the residual signal:

$$\varepsilon(k) = \int_a^b \sigma(z)(\sqrt{\gamma(z,F,k)} - \sqrt{\hat{\gamma}(z,k)})\mathrm{d}z \tag{11.9}$$

where $\sqrt{\hat{\gamma}(z,k)} = B(z)E\hat{x}(k) + h(E\hat{x}(k))B_n(z)$. In this case, the nonlinear fault detection filter can be formulated as

$$\hat{x}(k+1) = A\hat{x}(k) + A_d\hat{x}(k-d) + Gg(\hat{x}(k)) + Hu(k) + L\varepsilon(k) \tag{11.10}$$

where $\hat{x}(k)$ is the estimated state, $L \in R^{m\times p}$ is the gain to be determined and $\sigma(z) \in R^{p\times 1}$ can be regarded as a pre-specified weighting vector defined on $[a,b]$.

Denote $\tilde{h}(k) = h(Ex(k)) - h(E\hat{x}(k))$ and $\tilde{g}(k) = g(x(k)) - g(\hat{x}(k))$ as two auxiliary vectors, and $X(k) = [\mathrm{e}^T(k), \mathrm{e}^T(k-d), \tilde{h}^T(k), \tilde{g}^T(k)]^T$ as an augmented vector. The estimation error system for $e(k) = x(k) - \hat{x}(k)$ can be simply described by

$$e(k+1) = \tilde{A}X(k) + DF(k) \tag{11.11}$$

where $\tilde{A} = [A - L\Gamma_1, A_d, -L\Gamma_2, G]$ is an auxiliary state matrix.

For simplicity, it is assumed that $\hat{x}_j = 0, j = -d, -d+1, \cdots, -1, 0$. Thus we have $e(j) := \varphi(j) - \hat{x}(j), j = -d, -d+1, \cdots, -1, 0$. Denote the auxiliary matrices as

$$\begin{cases} \Gamma_1 = \int_a^b \sigma(z)B(z)E\mathrm{d}z, \ \Gamma_2 = \int_a^b \sigma(z)b_n(z)\mathrm{d}z \\ \Pi_1 = \mathrm{diag}\{\pi_0, -Q, -\lambda_1^2 I + \Gamma_2^T\Gamma_2, -\lambda_2^2 I\} \\ \Pi_0 = -P + Q + \lambda_1^2 E^T U_1^T U_1 E + \lambda_2^2 U_2^T U_2 + \Gamma_1^T\Gamma_1 \\ \Pi_2 = [PA - R\Gamma_1, PA_d, -R\Gamma_2, PG] \end{cases} . \tag{11.12}$$

Residual $\varepsilon(k)$ represents the difference of $\hat{\gamma}(z,k)$ and $\gamma(z,F,k)$, which can be formulated as the following nonlinear function of $X(k)$:

$$\varepsilon(k) = \Gamma_1 e(k) + \Gamma_2\tilde{h}(k). \tag{11.13}$$

*Remark 11.2* In [66], for continuous-time systems without time delays, we have provided sufficient conditions for the boundedness of the error system in the absence of the fault, which gives the threshold to detect $F(k)$. However, such an upper bound may be very conservative to detect $F(k)$, especially when the initial value is relatively large.

To this end, we construct the reference vector $z_1(k) = \varepsilon(k)$ and consider the sub-optimal guaranteed cost problem for $\min_L J_1 = \|z_1(k)\|_2^2$ subject to the error system

(11.11). The following problem will be studied: to find $L$ such that the occurrence of the fault can be detected using the response of $\varepsilon(k)$ as sensitively as possible. If the minimum $J_0$ for the estimation error system (11.11) in the absence of $F(k)$ can be found, the alarm of fault can be made when $\|\varepsilon(k)\|^2 \geq J_0$ holds. The following result is an extension of Lemma 1 in [57] from the control problem to the FDD filtering one.

**Theorem 11.1** If for the parameters $\lambda_i > 0 (i = 1,2)$, there exist matrices $P > 0$, $Q > 0$, and $R$ satisfying

$$\Pi = \begin{bmatrix} \Pi_1 & \Pi_2^T \\ \Pi_2 & -P \end{bmatrix} < 0, \tag{11.14}$$

then in the absence of $F$, error system (11.11) with gain $L = P^{-1}R$ is asymptotically stable and the error satisfies

$$J_1 \leq \alpha_1^2 := \mathrm{tr}(S_0^T P S_0) + \Sigma_{l=1}^d \mathrm{tr}(S_l^T Q S_l). \tag{11.15}$$

If the following optimal problem $\min_{P>0,Q>0,R,\eta>0} J_1 := \alpha_{min,1}$ is solvable subject to Equation 11.21, then fault $F(k)$ can be detected if $\|\varepsilon(k)\|_2 > \alpha_{min,1}$ holds.

*Proof.* Define the Lyapunov candidate as follows:

$$\Phi(k) := \mathrm{e}^T(k)Pe(k) + \sum_{i=k-d}^{k-1} \mathrm{e}^T(i)Qe(i) + \lambda_1^2 \sum_{i=1}^{k-1} [\|U_1 Ee(i)\|^2 - \|\tilde{h}(i)\|^2]$$

$$+\lambda_2^2 \sum_{i=1}^{k-1} [\|U_2 e(i)\|^2 - \|\tilde{g}(i)\|^2], \tag{11.16}$$

which can be guaranteed to be positive based on Equations 11.9 and 11.11. In the absence of $F(k)$, along with Equation 11.11 it can be verified that

$$\begin{aligned}\Delta\Phi(k) &= \Phi(k+1) - \Phi(k) \\ &= \mathrm{e}^T(k+1)Pe(k+1) - \mathrm{e}^T(k)Pe(k + \mathrm{e}^T(k)Qe(k) - \mathrm{e}^T(k-d)Qe(k-d) \\ &\quad +\lambda_1^2\|U_1Ee(k)\|^2 - \lambda_1^2\|\tilde{h}(k)\|^2 + \lambda_2^2\|U_2e(k)\|^2 - \lambda_2^2\|\tilde{g}(k)\|^2 \\ &= X_k^T \Psi X_k\end{aligned}$$

where

$$\Psi = \begin{bmatrix} \Psi_1 & (A - L\Gamma_1^T)PA_d & -(A - L\Gamma_1^T)PL\Gamma_2 & (A - L\Gamma_1^T)PG \\ * & -Q + A_d^T PA_d & -A_d^T PL\Gamma_2 & A_d^T PG \\ * & * & -\lambda_1^{-2}I + \Gamma_2^T LPL\Gamma_2 & -\Gamma_2^T LPG \\ * & * & * & -\lambda_2^{-2}I + G^T PG \end{bmatrix}$$

and $\Psi_1 = (A - L\Gamma_1^T)^T P(A - L\Gamma_1^T) - P + Q + \lambda_1^{-2}E^T U_1^T U_1 E + \lambda_2^{-2}U_2^T U_2$. Furthermore, we denote an auxiliary function as follows:

$$J_{1a} = \sum_{l=0}^{\infty} [z_1^T(l) z_1(l) + \Delta \Phi(l)]. \tag{11.17}$$

Similarly to Equation 11.16, it can be verified that $J_{1a} \leq X_k^T \tilde{\Psi} X_k$, where $\tilde{\Psi} = \Psi + \mathrm{diag}\{\Gamma_1^T \Gamma_1 + \Gamma_2^T \Gamma_2, 0\}$.

Based on the Schur complement formula, it is easy to see that $\tilde{\Psi} < 0$ is equivalent to $\Pi < 0$. This means that error system (11.11) is asymptotically stable under $\Pi < 0$. Because $\lim_{k\to\infty} e(k) = 0$, Equations 11.5 and 11.7 guarantee that $\lim_{k\to\infty} \tilde{h}(k) = 0$ and $\lim_{k\to\infty} \tilde{g}(k) = 0$, which implies that $\lim_{k\to\infty} \Phi(k) = 0$. Thus, it can be verified that $J_{1a} = J_1 + \lim_{k\to\infty} \Phi(k) - \Phi(0) = J_1 - \Phi(0)$ and consequently $J_1 \leq J_{1a} + \tilde{a}_1^2$ by using the definition of $J_{1a}$ and $\tilde{a}_1^2 := \varphi^T(0) P \varphi(0) + \sum_{l=1}^{d} \varphi^T(-l) P \varphi(-l)$. On the other hand, $J_{1a} \leq 0$ holds because $J_{1a} \leq X_k^T \tilde{\Psi} X_k$ and $\tilde{\Psi} < 0$. Thus, Equation 11.15 can be obtained based on the definition of $S$.

From Equation 11.16 and based on Theorem 11.1, when $F = 0$, it can be shown that $\|\varepsilon(k)\| \leq \alpha_{1min}$ holds. This implies that the residual vector $\varepsilon(k)$ should satisfy $\|\varepsilon(k)\| > \alpha_{1min}$ in the presence of $F$. □

Theorem 11.1 provides an LMI-based stability criterion for the error system in the absence of the fault, where the threshold is also determined to be $\alpha_{min,1}$.

## 11.4 Fault Diagnosis Filter Design

In this section, we consider a fault diagnostic filter to estimate the size of the fault using the measured output PDFs. In this context, the following adaptive observer can be constructed:

$$\begin{cases} \hat{x}(k+1) = A\hat{x}(k) + A_d \hat{x}(k-d) + Gg(\hat{x}(k)) + Hu(k) + L\varepsilon(k) + D\hat{F}(k) \\ \hat{F}(k+1) = \hat{F}(k) - \Upsilon \varepsilon(k) \end{cases} \tag{11.18}$$

where $\hat{F}(k)$ is the estimation of $F(k)$, $\Upsilon$ is a learning operator which will be determined by the following diagnostic method together with $L$. Regarding the second equation of (11.18), it can be shown that it not only an observer to estimate $F(k)$, but also an iterative learning control law acting on the first equation with an undetermined learning operator. The information for feedback as the new type of residual can also be described by Equation 11.13.

Denote $e(k) = x(k) - \hat{x}(k)$, $f(k) = F(k) = \hat{F}(k)$. For shortened descriptions, we adopt the following notation for the coefficient matrices:

$$\bar{A} = \begin{bmatrix} A & D \\ 0 & I \end{bmatrix}, \bar{A}_d = \begin{bmatrix} A_d & 0 \\ 0 & 0 \end{bmatrix}, \bar{\Gamma}_1^T = \begin{bmatrix} \Gamma_1^T \\ 0 \end{bmatrix}, \bar{G}^T = \begin{bmatrix} G \\ 0 \end{bmatrix}, \bar{L} = \begin{bmatrix} L \\ \Upsilon \end{bmatrix}$$

and $\xi^T(k) = [e^T(k), f^T(k)]^T$ for the augmented variable. Thus, the error system can be simplified to

$$\xi(k+1) = \bar{A}_c x(k) + \bar{A}_d x(k-d) + \bar{G}\tilde{g}(k) + \bar{L}\bar{\Gamma}_2\tilde{h}(k) \tag{11.19}$$

where $\bar{A}_c = \bar{A} + \bar{L}\bar{\Gamma}_2$ . It can verified that $\|\tilde{h}(k)\| \leq \|\bar{U}_1\bar{E}\xi(k)\|$, $\|\tilde{g}(k)\| \leq \|\bar{U}_2\xi(k)\|$, where $\bar{U}_1 = \mathrm{diag}\{U_1, 0\}$, $\bar{U}_2 = \mathrm{diag}\{U_2, 0\}$ and $\bar{E} = \mathrm{diag}\{E, 0\}$.

The first objective is to find $L$ and $\Upsilon$ (or $\bar{L}$) such that system (11.19) is asymptotically stable, which can guarantee that the fault can be asymptotically tracked. However, such a criterion is conservative since, practically, the tracking performance has to be achieved in a limited time with a satisfactory estimated range. A guaranteed cost function as a type of performance index has been widely used in robust control, where the norm of a reference output can be guaranteed within a pre-specified bound (see [57, 203] and references therein). However, even for the transformed fault diagnosis problem of nonlinear discrete-time systems with time delays, few results have been given due to its complexity. In this chapter by establishing the above observer with an ILC structure, we will give an algorithm to design the gains such that the error system is both stable and with a guaranteed cost performance with respect to a reference output. To meet the requirement of fault diagnosis, the reference output can be selected as

$$z_2(k) = W\xi(k) = W_1 e(k) + W_2 f(k)$$

where $W$ (or $W_1$ and $W_2$) are given weighting matrices. It can be supposed that $F(k) = 0$ and $\hat{F}(k) = 0$ at the initial time, which corresponds to $f(k) = 0$ and $\xi(k) = [\varphi_k^T, 0]^T$ when $d \leq k \leq 0$.

In [66], an observer-based fault diagnosis approach has also been given for continuous-time systems, where only the boundedness of the estimation error can be guaranteed. The following result improves the performance of the diagnosis filter, where $J_2 = \|z_2(k)\|_2^2$ can be minimized or confined to a pre-specified range.

**Theorem 11.2** For the parameters $\lambda_i (i = 1,2)$ , if there exist matrices $P = \begin{bmatrix} P_{11} & P_{12} \\ P_{12}^T & P_{22} \end{bmatrix} > 0$ , $Q = \begin{bmatrix} Q_{11} & Q_{12} \\ Q_{12}^T & Q_{22} \end{bmatrix} > 0$ , and $R$ satisfying

$$\Omega = \begin{bmatrix} \Omega_1 & 0 & 0 & 0 & \bar{A}^T P + \bar{\Gamma}_1^T R^T \\ 0 & -Q & 0 & 0 & \bar{A}_d^T P \\ 0 & 0 & -\lambda_1^2 I & 0 & \bar{G}^T P \\ 0 & 0 & 0 & -\lambda_2^2 I & \bar{\Gamma}_2^T R^T \\ P\bar{A} + R\bar{\Gamma}_1 & P\bar{A}_d & P\bar{G} & R\bar{\Gamma}_2 & -P \end{bmatrix} < 0 \tag{11.20}$$

where $\Omega_1 = Q + \lambda_1^2 \bar{E}^T \bar{U}_1^T \bar{U}_1 \bar{E} + \lambda_2^2 \bar{U}_2^T \bar{U}_2 - P$, then using diagnostic observer (11.18) with gain $\bar{L} = P^{-1}R$, the error system (11.19) is asymptotically stable and the error satisfies

$$J_2 \leq \alpha_2^2 := \mathrm{tr}(S_0^T P_{11} S_0) + \sum_{l=1}^{d} \mathrm{tr}(S_l^T Q_1 S_l). \tag{11.21}$$

*Proof.* Define the following Lyapunov function:

$$\Theta(k) := \xi^T(k)P\xi(k) + \sum_{i=k-d}^{k-1} \xi^T(i)Q\xi(i) + \lambda_1^2 \sum_{i=1}^{k-1} [\|U_1Ee(i)\|^2 - \|\tilde{h}(i)\|^2]$$
$$+\lambda_2^2 \sum_{i=1}^{k-1} [\|U_2e(i)\|^2 - \|\tilde{g}(i)\|^2].$$

Along the trajectories of the error system, it can be obtained that

$$\Delta\Theta(k) = \Theta(k+1) - \Theta(k)$$
$$= \xi^T(k+1)P\xi(k+1) - \xi^T(k)P\xi(k) + \xi^T(k)Q\xi(k) - \xi^T(k-d)Q\xi(k-d)$$
$$+\lambda_1^2\|U_1Ee(k)\|^2 - \lambda_1^2\|\tilde{h}(k)\|^2 + \lambda_2^2\|U_2e(k)\|^2 - \lambda_2^2\|\tilde{g}(k)\|^2.$$

It can be seen that

$$\xi^T(k+1)P\xi(k+1) = Y^T(k)\Xi_1 Y(k)$$

where $Y(k) = [\xi^T(k), \xi^T(k-d), \tilde{g}^T(k), \tilde{h}^T(k)]^T$ and

$$\Xi_1 = [\hat{A}_c, \hat{A}_d, G, S\Gamma_2]^T P [\hat{A}_c, \hat{A}_d, G, S\Gamma_2].$$

Define $\Xi_2 = \mathrm{diag}\{\Omega_1, -Q, -\lambda_1^2 I, -\lambda_2^2 I\}$, where

$$\Omega_1 = Q + \lambda_1^2 \bar{E}^T \bar{U}_1^T \bar{U}_1 \bar{E} + \lambda_2^2 \bar{U}_2^T \bar{U}_2.$$

It can be seen that $\Delta\Theta(k) = Y^T(k)[\Xi_1 + \Xi_2]Y(k)$. Define the following auxiliary function:

$$J_2 = \sum_{l=0}^{\infty} [z_2^T(l)z_2(l) + \Delta\Theta(l)].$$

Furthermore, it can be seen that

$$z_2^T(k)z_2(k) = \xi^T(k)W^T W\xi(k). \tag{11.22}$$

Thus, we can obtain

$$z_2^T(k)z_2(k) + \Delta\Theta(k) = \xi^T(k)W^T W\xi(k) + Y^T(k)[\Xi_1 + \Xi_2]Y(k) = Y^T(k)\Xi Y(k) \tag{11.23}$$

where $\Xi = \mathrm{diag}\{W^T W, 0, 0, 0\} + \Xi_1 + \Xi_2$. As a result, using the Schur complement formula and algebraic transformations, it is shown that $\Xi < 0$ holds if and only if

$$\begin{bmatrix} \Omega_1 + W^T W & 0 & 0 & 0 & \bar{A}^T P + \bar{\Gamma}_1^T S^T P \\ * & -Q & 0 & 0 & \bar{A}_d^T P \\ * & * & -\lambda_1^2 I & 0 & \bar{G}^T P \\ * & * & * & -\lambda_2^2 I & \bar{\Gamma}_2^T S^T P \\ * & * & * & * & -P \end{bmatrix} < 0, \tag{11.24}$$

which is equivalent to $\Omega < 0$ with $R = PL$.

Under $\Omega < 0$ and inequality (11.24), first we can see that $\Xi_1 + \Xi_2 < 0$, which implies that the error system is asymptotically stable. Next we have $J_{2a} = J_2 - \Theta(0) < 0$, which leads to $J_2 < \tilde{a}_2^2$, where $\tilde{a}_2^2 = \xi^T(0)P\xi(0) + \Sigma_{l=1}^d \xi^T(-l)P\xi(-l)$. Equation 11.19 can be verified by using simple algebraic transformations. □

*Remark 11.3* In Theorem 11.2, not only the observer gain, but also the learning operator can be designed using the solutions of LMIs. Discrete-time models have been widely used in digital control engineering. However, analysis and synthesis for discrete-time system may have additional complexity. Compared with the corresponding results for continuous time systems (see [66]), it can be proved that a similar analysis approach, where the Lyapunov function has been chosen as a diagonal matrix, will be unavailable for the fault diagnosis problem considered.

## 11.5 Simulation Examples

Similarly to [66], we suppose that these output PDFs can be approximated using square root B-spline models described by Equation 11.1, with $n = 3$, $z \in [0, 1.5]$ and

$$b_i = \begin{cases} |\sin 2\pi z|, & z \in [0.5(i-1),\ 0.5i] \\ 0, & others \end{cases} \quad (i = 1, 2),$$

$$b_3 = \begin{cases} \frac{1}{10}|\sin 2\pi z|, & z \in [1,\ 1.5] \\ 0, & others \end{cases}.$$

It is assumed that the identified weighting system is formulated by Equation 11.6 with the following coefficient matrices:

$$A = \begin{bmatrix} -0.5 & 0.01 \\ 0.125 & -0.31 \end{bmatrix}, A_d = \begin{bmatrix} 0.1 & 0 \\ 0 & 0.1 \end{bmatrix}, H = \begin{bmatrix} 0.5 & 0 \\ 0 & -0.9 \end{bmatrix},$$

$$G = \begin{bmatrix} 0 & 0.2 \\ 0 & 0.2 \end{bmatrix}, E = \begin{bmatrix} 0.101 & 0.1001 \\ 0.010 & 0.001 \end{bmatrix}, J = \begin{bmatrix} 1.33 \\ -0.1 \end{bmatrix}.$$

The bound of uncertainties is denoted by $U_1 = \mathrm{diag}\{0.1, 0.1\}$ and $U_2 = \mathrm{diag}\{0, 0.5\}$. From Equation 11.3, it can be seen that $\Lambda_1 = \mathrm{diag}\{0.25, 0.25\}$, $\Lambda_2 = [0\ \ 0]$, $\Lambda_3 = 0.0025$. It can be obtained that $\Gamma_1 = [0.1953\ \ 0.0108]$, $\Gamma_2 = 0.0971$ for $\sigma(z) = 3.051$. In simulations, we select $x(k) = [0.1 + e^(k-5)\ \ -0.1 + e^(k-5)]^T$ $(k \in [-5, 0])$, $\hat{x}(k) = [0\ \ 0]^T$ $(k \in [-5, 0])$, and $\lambda_1 = \lambda_2 = 1$.

When the fault is selected as $F(k) = \begin{cases} 0.6, & k \geq 200 \\ 0, & k < 200 \end{cases}$, it can be shown that the corresponding result of [66] is unfeasible to detect $F(k)$. Using Theorem 11.2, it can be obtained that

$$P = \begin{bmatrix} 2.9244 & -0.0306 \\ -0.0306 & 3.1712 \end{bmatrix}, Q = \begin{bmatrix} 1.0546 & 0.0040 \\ 0.0040 & 0.9344 \end{bmatrix},$$

$$R = \begin{bmatrix} -6.5117 \\ 1.5048 \end{bmatrix}, L = \begin{bmatrix} -2.2219 \\ 0.4531 \end{bmatrix}.$$

The residual signal should satisfy $|\varepsilon(t)| > \alpha_1 = 0.1206$, where $\alpha_1$ is the threshold determined by Equation 11.15. Figure 11.1 shows the responses of the residual and the threshold. It is noted that to show robustness, in the simulation, we added a random noise as the modeling error acting on Equation 11.8, which is selected as a random input within $[-0.0366, 0.0366]$.

For the diagnosis problem, using LMI tools, the following corresponding results can be obtained:

$$P = \begin{bmatrix} 0.0853 & 0.9988 \\ 0.9988 & 13.3537 \end{bmatrix}, Q = \begin{bmatrix} 0.0412 & 0.4667 \\ 0.4667 & 6.1247 \end{bmatrix},$$

$$R = \begin{bmatrix} 0.5219 \\ 7.6334 \end{bmatrix}, L = \begin{bmatrix} -4.6310 \\ 0.9180 \end{bmatrix}, \Upsilon = 0.0267.$$

We assume $F(k) = 0.6$ occurs at the 200th second. The error response is shown in Figure 11.2. In Figure 11.3 the fault and its estimate are given. It is shown that the tracking performance is achieved. Furthermore, we consider the time-varying fault described by

$$F(k) = \begin{cases} 10 + 0.15\sin(0.01k), & k \geq 200 \\ 0, & k < 200 \end{cases}$$

and investigate the influence of the modeling error. The modeling error is selected as a random input within $[-0.0366, 0.0366]$. Simulations show that the fault can also be well estimated using the designed diagnostic filter. Figure 11.4 shows the convergence of errors and Figure 11.5 shows the fault and its estimated value using the diagnostic filter. Although theoretically the result is only for a constant fault in the absence of modeling error, this example implies that this result is robust for some time-varying faults in the presence of modeling errors.

## 11.6 Conclusions

This chapter considers a new type of discrete-time FDD problem that originated from some practical processes, along with the development of instruments and image processing techniques. Different from conventional FDD problems, the measurement information for FDD is the output stochastic distributions and the stochastic variables involved are not confined to Gaussian ones. With the square root B-spline expansions the new type of FDD problem can be formulated as a classical FDD problem subject to a discrete-time nonlinear time-delayed system. Noting that few FDD approaches have been given even for the transformed model, we provide a new fault detection and fault diagnosis observer design scheme, where guaranteed cost performance optimization is used to improve the capability of the proposed FDD observers. Simulations demonstrate the effectiveness and advantages of the

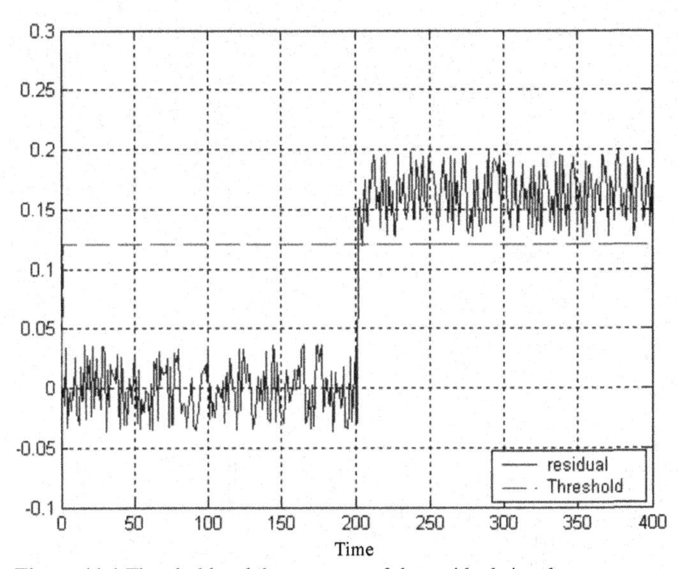


**Figure 11.1** Threshold and the response of the residual signal

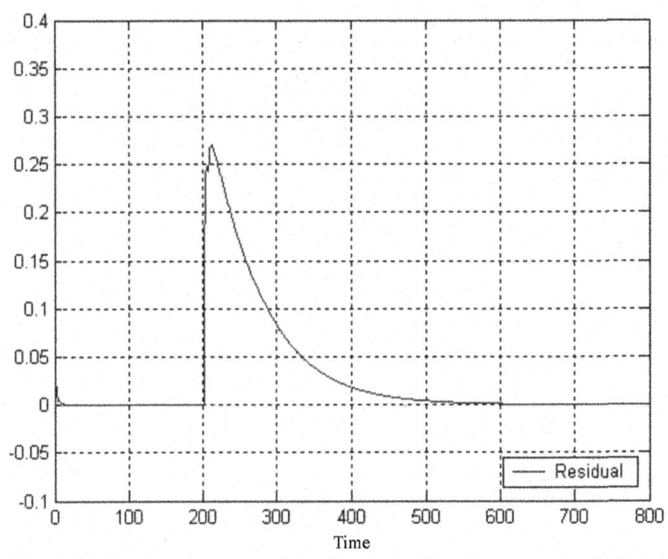


**Figure 11.2** Estimation error of the diagnostic filter for a constant fault

proposed FDD approaches. Further investigations are required on systems with a time-varying fault and modeling errors.

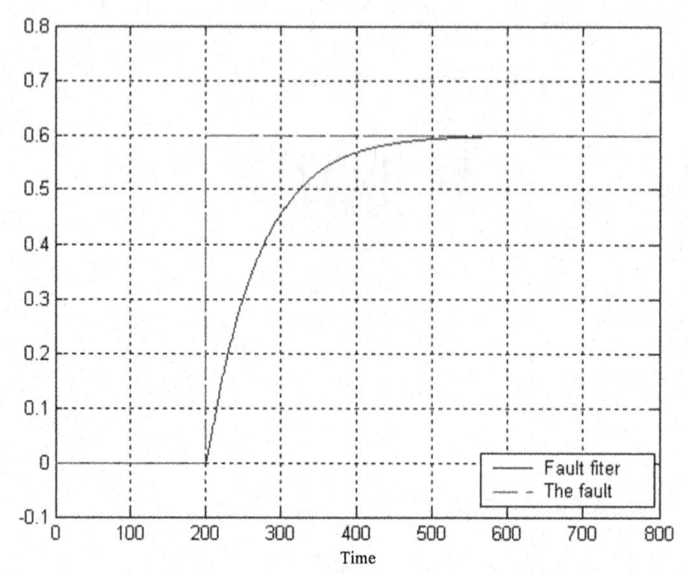


**Figure 11.3** Response of the diagnostic filter and the constant fault

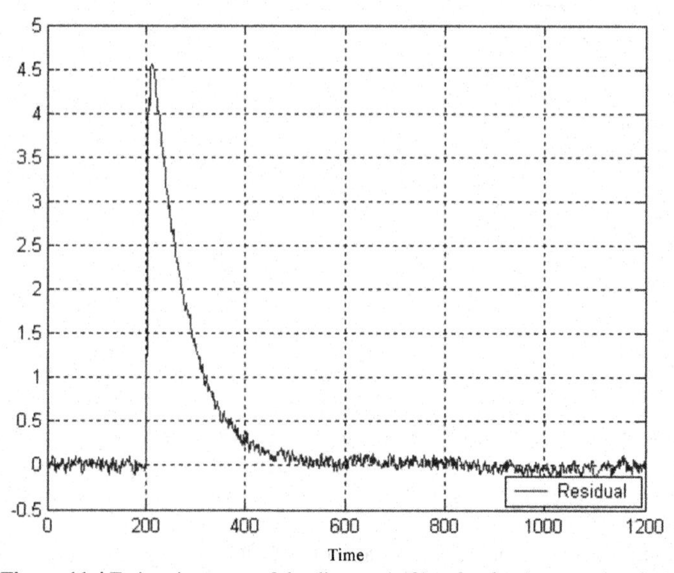


**Figure 11.4** Estimation error of the diagnostic filter for the time-varying fault

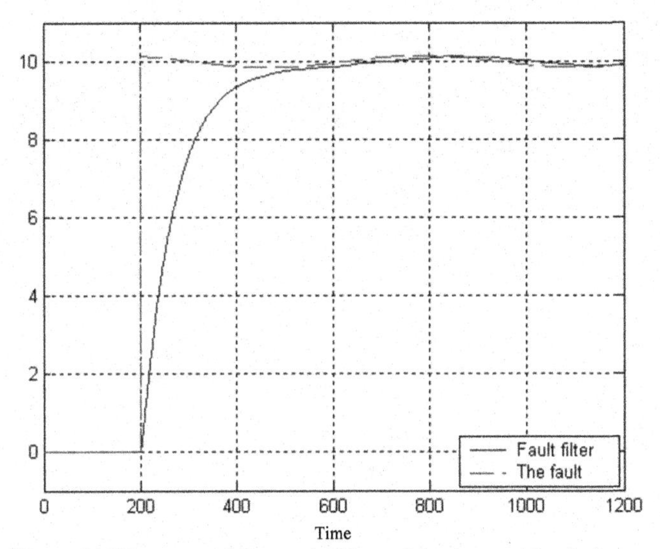


**Figure 11.5** Response of diagnostic filter and the time-varying fault

# Part V
# Conclusions

# Chapter 12
# Summary and Potential Applications

## 12.1 Summary

After twelve years of development, research into stochastic distribution control has gone through a number of stages from the early B-spline-based control design to the recent input and output model-based design for general stochastic systems. During these developments, many novel techniques such as LMI-based convex optimization, NN and fuzzy dynamic modeling, structural and adaptive tracking control, and model-based FDD have been developed and widely used to formulate controllers together with effective closed-loop analysis. This book has established a unified framework on SDC, where after the introduction in Chapter 1, in the following chapters we focused on the structural controller design, two-step intelligent optimization modeling and control for SDC systems, statistic tracking control driven by output statistical information, and FDD for SDC systems.

Compared with previous work, the results in the book have several novel features. From the perspective of modeling, generalized two-step NN models are established for nonlinear and non-Gaussian systems such that LMI-based optimization tools can be used to control the shape of the output PDFs. In terms of closed-loop system analysis, global stability, robustness, dynamic tracking performance can be assessed systematically. As for the control design, a feasible LMI-based convex optimization framework has been established to solve the multiple objective control problems related to PDF tracking and statistical information tracking. Finally, from the perspective of filtering and FDD, this book has, for the first time, provided a feasible way to solve the filtering and FDD problem for non-Gaussian systems.

The advantage of using LMI is that the optimization for PDF tracking can be effectively transferred into a convex optimization problem, where a global minimum effect can be achieved. This differs from previous developments where numerical optimization was used, which only guarantees a local minimum. Also, closed-loop system stability analysis can be carried out in a much simpler and rigorous framework. In addition, the two-stage modeling of SDC systems constitutes another outstanding feature in terms of combined use of data-based modeling and LMI-based analysis.

Moreover, the theoretical results described in this book have also enriched the development of LMI-based control and estimation theory. The main contributions include the generalized PI/PID controller based on convex LMI algorithms designed to achieve the tracking control problem of MIMO nonlinear systems; the multiple objective control problem (including stability, tracking performance, robustness and constraint) for complex nonlinear systems has been solved through the designed convex optimization algorithms. Also, the new nonlinear FDD and FTC methodologies have also been developed via observer or filter design theory, and many of them have been applied to practical processes successfully.

## 12.2 Potential Applications of PDF Control

The idea of PDF control has several potential applications in data-driven modeling, data reduction and plant-wide optimization of complex industrial processes. These issues will be briefly described in this section.

### *12.2.1 PDF Control in Data-driven Modeling*

Another application of PDF control is for data-driven modeling of unknown systems. It is well known that when input and output data are used to establish models for unknown systems, a model with a set of tuning parameters (such as NN weights) is used, where the input to the model is the actual input data sequence of the system. This input sequence works together with the tuning parameters of the model to generate a model output. In this context, the modeling error, or residual, is defined as the difference between the actual system output and the model output. A good model would be such that this modeling error is made as small as possible. When there is noise contained in the original input and output data collected from the unknown system, the variance of the modeling error has been widely used as a measure to see whether the model obtained is good or not. As such, during the modeling phase, these tuning parameters are selected so that the variance of the modeling error is minimized and hence the mean value of the modeling error is close to zero. If the modeling error is Gaussian, such minimum variance-based modeling would mean that the shape of the PDF of the modeling error is made as narrow as possible. However, in most cases, the PDF of the modeling error is not Gaussian. Therefore, the tuning parameters embedded in the model should be selected so that the PDF of the modeling error is made as close as possible to a narrowly distributed Gaussian PDF (see Figure 12.1). This is an output PDF control problem, where the control input is the set of tuning parameters and the output is the PDF of the modeling error. Of course, the tuning parameters can also be selected so that the entropy of the modeling error is minimized. As a result, it can be concluded that output PDF control can find potential applications in data-driven modeling of unknown systems.

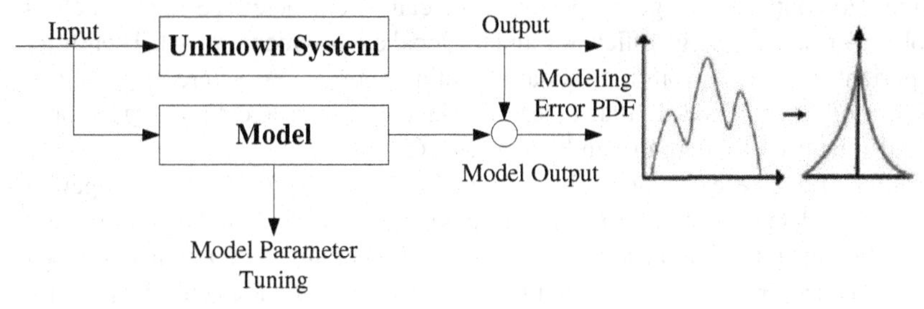


**Figure 12.1** Data-based modeling as PDF shape control

### *12.2.2 PDF Control in Data Dimension Reduction*

Principal component analysis (PCA) has been widely used to perform high dimensional data reduction, where a large volume of data (normally as a result of sampling a few thousands of variables) can be reduced to a small number of principal components that contain most of the information of the original high dimensional data. Once these principal components are obtained, they can be used to (approximately) recover the original data through inverse mapping. Normally, good data reduction would mean that the recovery error is minimum variance with zero mean, which is based on Gaussian assumptions. For a non-Gaussian recovery error, one needs to select the principal components so that the PDF of the recovery error is made as close as possible to a narrowly distributed Gaussian PDF. This is again an output PDF control problem where the control input is the set of principal components and the output is the PDF of the recovery error.

### *12.2.3 PDF Control in Multi-scale Plant-wide Modeling for Fault Diagnosis and Fault Tolerant Control*

As previously described, data-driven models should be considered because they have the advantage of exploiting the large volume of process data that exist in distributed control systems (DCS) found in many process industries. Once these data-driven models are established, they can be used for a number of purposes including plant-wide optimization, plant-wide performance monitoring and, of course, FDD and FTC.

Complex industrial processes are examples of sequentially arranged multi-agent systems where each production unit is connected in series so as to process the required intermediate products. Each product unit can be regarded as an agent and each agent would have a number of control systems so as to make it autonomous. Each agent has its own performance that needs to be considered in plant-wide optimization and control. The agent's performance can be characterized in terms of its control performance, energy efficiency and the local environment impact. Therefore, an important difference from traditional modeling approaches, where only control function models are used, is that other criteria such as an agent's energy model should also be used for data-driven FDD and FTC.

Potential future research on multi-scale modeling could focus on investigating where process data should be best used for modeling sequentially linked multi-agent systems. For example, along a general material processing production line, each unit (agent) accepts the material to be processed from the previous unit. Its purpose is to process the material so that it will have the required properties after passing through the unit. One can therefore represent the vector of material properties as $X_n$ and $X_{n+1}$, prior to entering and after leaving the nth unit, respectively. Then the following structure represents the functioning of the nth unit:

$$X_n \otimes U_n(t_n, t_{n+1}) \rightarrow X_{n+1} \tag{12.1}$$

where $U_n(t_n, t_{n+1})$ are the control variables applied in the nth unit between time $t_n$ and $t_{n+1}$ so as to transform the material properties from $X_n$ into $X_{n+1}$, and $\Delta t_n = t_{n+1} - t_n$ is the time duration for the material to pass through the nth unit. Although different plants will have different models for each of the sequentially arranged units, the above structure can be generally applied as a signal pathway representation of material flow. In this context, the relationship between $(X_n, U_n)$ and $X_{n+1}$ can be represented as dynamic and possibly stochastic equations which are to be determined.

In Equation 1.8, $X_n$ has components represented either by time-dependent variables (such as materials strength, etc.) or by variable distributions (such as the particle size distributions represented by a PDF in polymerization processes). Moreover, the components in $X_n$ can be classified into two groups, namely measurable and un-measurable components. Since $U_n(t_n, t_{n+1})$ can be either the control inputs to each unit or the set-point values from the DCS top layer, the models obtained will be multi-layer in structure. In addition, the communication between each unit and across each layer should also be considered in the modeling phase. The model therefore has a multi-layer structure in terms of control loops in each unit and the material properties transformations between $X_n$ and $X_{n+1}$. As all the units are connected in series, the time delay of each unit with respect to the final products needs to be taken into account. Specifically, the reference time can be selected as the time when the products are produced. Due to the variations in the processing duration when the materials pass through each unit, these time delays will be time-varying in nature. It is expected that a general multi-scale model structure, which takes into account process dynamics, will be developed for each agent as well as the whole system.

Using the models (12.1), one of the future areas of research is to develop global PDF models that link relevant control variables in each production unit with the

PDF of the final quality variables. For this purpose, a back propagation modeling procedure will be developed that links various unit models together via the time delay and process inherent multi-layer connections so as to form the following global and conditional PDF model for production qualities:

$$\gamma(y, U_k) = \gamma(y, U_N(t_{N-1}, t_N), U_{N-1}(t_{N-2}, t_{N-1}), \cdots, U_1(t_0, t_1)) \tag{12.2}$$

where $U_k$ denotes all the controllable variables of the system. When the sectional control variables are high dimensional, PCA algorithms can be used for data reduction so as to reduce the dimensionality of $U_k$. Therefore, the final control vector that appears in the quality variables' PDFs will be either in terms of the original variables or in a projected subspace. Probability theory can also be applied to each sectional model to develop the quality variables' PDFs, where the recently developed recursive PDFs model can also be used. Indeed, research into such data-driven multi-scale models would therefore constitute a potential starting point for data-driven FDD and FTC.

In line with the multi-scale models constructed for industrial processes, FDD and FTC (either model-based or data-based) can also be developed to deal with the faults for individual agents. However, a possible new research direction would be to investigate the quality data distribution of each agent or the final production quality. It is well known that product quality monitoring has been used for many years, however, the existing techniques are generally based on monitoring the means and the variances of important product quality variables in the data stream. This assumes that the process quality data is Gaussian distributed, however, for complex industrial processes, the quality data distribution is generally non-Gaussian. Often, it is only during healthy operation that the quality data distribution is approximately Gaussian.

This phenomenon provides a potential insight toward the realization of data-based FDD and FTC for complex industrial plant. For example, when the quality data PDF is not Gaussian, then it may indicate that there is either a fault in the process or a variation in the raw material. As such, the purpose of fault diagnosis is to first classify whether the non-Gaussian PDF is caused by material variations or by a process fault. Once it has been confirmed that the non-Gaussian PDF is caused by the process fault, further fault diagnosis analysis is performed to locate the fault and estimate its size. In this context, observer-based fault diagnosis can be used, where PDF residuals are constructed for the sequential PDF models. Adaptive tuning rule-based fault diagnosis methods can be developed so as to guarantee the performance of the fault diagnosis. Once the fault diagnosis results are obtained, they will be combined with the relevant FTC control algorithm so that the whole plant-wide control strategy will be fault tolerant. Moreover, FTC design can also be made to minimize the entropy of the quality data so that a consistent product quality with minimum uncertainty (randomness) can be achieved.

# References


[1] Anderson BDO, Moore JB. Linear optimal control. Prentice Hall, Inc., 1971.

[2] Ang KH, Chong GCY, Li Y. PID control system analysis, design and technology. IEEE Trans Control Syst Technol. 2005; 13(4): 559-76.

[3] Åström KJ. Introduction to stochastic control theory. New York, Academic, 1970.

[4] Åström KJ, Hagglund T. PID Controllers: theory, design, and tuning. 2nd ed. Research Triangle Park, NC, Instrument Society of America, 1994.

[5] Åström KJ, Wittenmark B. Adaptive control. Reading, MA, Addison-Wesley, 1989.

[6] Basseville M, Nikiforov I. Fault isolation for diagnosis: nuisance rejection and multiple hypothesis testing. Annu Rev Control. 2002; 26(2): 189-202.

[7] Boukas E, Liu ZK. Suboptimal design of regulators for jump linear systems with time-multipled quadratic cost. IEEE Trans Automatic Control 2001; 46(1): 131-6.

[8] Boyd S, Ghaoui LE, Feron E, Balakrishnam V. Linear matrix inequalities in system and control theory. Philadelphia, SIAM Studies in Applied Mathematics, 1994.

[9] Braatz RD. Advanced control of crystallization processes. Annu Rev Control. 2002; 26(1): 87-99.

[10] Brown M, Harris CJ. Neurofuzzy adaptive modeling and control. Englewood Cliffs, NJ, Prentice-Hall, 1994.

[11] Cameron I, Wang FY, Immanuel CD, Stepanek F. Process systems modelling and applications in granulation: a review. Chem Eng Sci. 2005; 60(14), 3723-50.

[12] Campbell GM, Bunn PJ, Webb C, Hook SCW. On predicting roll milling performance, Part II: the breakage equation. Powder Technol. 2001; 115(3): 234-55.

[13] Cao YY, Frank PM. Analysis and synthesis of nonlinear time-delay systems via fuzzy control approach. IEEE Trans Fuzzy Syst. 2000; 8(2): 200-11.

[14] Cerrillo JF, MacGregor JF. Control of particle size distributions in emulsion semibatch polymerization using mid-course correction polices. I&EC Res. 2002, 41(7), 1805-14.

[15] Cerrillo JF, MacGregor JF. Iterative learning control for final batch product quality using partial least squares models. I&EC Res. 2005; 44(24): 9146-55.

[16] Chairez I, Poznyak A, Poznyak T. New sliding-mode learning law for dynamic neural network observer. IEEE Trans Circuits Syst. II 2006; 53(12): 1338-42.

[17] Chen B, Liu XP, Tong SC, Ling C. Guaranteed cost control of T-S fuzzy systems with state and input delays. Fuzzy Sets Syst. 2007; 158(20): 2251-67.

[18] Chen HF, Guo L. Identification and stochastic adaptive control. Boston, Birkhiuser, 1991.

[19] Chen J, Patton R. Robust model-based fault diagnosis for dynamic systems. Kluwer Academic Publishers, 1999.

[20] Chen RH, Mingori DL, Speyer JL. Optimal stochastic fault detection filter. Automatica. 2003; 39(3): 377-90.

[21] Chou CH, Cheng CC. A decentralized model reference adaptive variable structure controller for large-scale time-varying delay systems. IEEE Trans Automatic Control. 2003; 48(7): 1213-17.

[22] Christofides PD. Control of nonlinear distributed parameter systems: recent developments and challenges. AIChE J. 2001; 47(3), 514-18.

[23] Clarke-Pringle TL, MacGregor JF. Optimization of molecular-weight distribution using both-to-batch adjustments. I&EC Res. 1998; 37: 3660-3669.

[24] Crespo LG, Sun JQ. Nonlinear stochastic control via stationary probability density functions. In Proc of the ACC 2002; 2029-34.

[25] Crespo LG, Sun JQ. Non-linear stochastic control via stationary response design. Probabilistic Eng Mech. 2003; 18(1): 79-86.

[26] Crowley TJ, Choi KY. Calculation of molecular weight distribution from molecular weight moments in free radical polymerization. I&EC Res. 1997; 36(5): 1419-23.

[27] Crowley TJ, Choi KY. Discrete optimal control of molecular weight distribution in a batch free radical polymerization process. I&EC Res. 1997; 36(9): 3676-84.

[28] Crowley TJ, Harrison CA, Doyle III FJ. Batch-to-batch optimization of PSD in emulsion polymerization using a hybrid model. In Proc of the ACC 2001, 981-6.

[29] Crowley TJ, Meadows ES, Kostoulas E, Doyle III TJ. Control of particle size distribution described by a population balance model of semibatch emulsion polymerization. J Process Control. 2000; 10(5): 419-32.

[30] Daoutidis P, Henson MA. Dynamics and control of cell populations in continuous bioreactor. In AIChE Symposium Series 2002, 326: 274-89.

[31] Dodson C, Wang H. Iterative approximation of statistical distributions and relation to information geometry. J Stat Inference Stoch Process. 2001; 4(3): 307-18.

[32] Eaton JW, Rawlings JB. Feedback control of chemical processes using on-line optimization techniques. Comput Chem Eng. 1990; 14: 469-79.

[33] Eek RA, Bosgra OH. Controllability of particulate processes in relation to the sensor characteristics. Powder Technol. 2000; 108(2): 137-46.

[34] Elbeyli O, Hong L, Sun JQ. On the feedback control of stochastic systems tracking pre-specified probability density functions. Trans Inst Measure Control. 2005; 27(5): 319-30.

[35] Fang HJ, Ye H, Zhong MY. Fault diagnosis of networked control systems. Annu Rev Control. 2007; 31: 55-68.

[36] Felix RA, Sanchez EN, Chen GR. Reproducing chaos by variable structure recurrent neural networks. IEEE Trans Neural Netw 2004; 15(6): 1450-57.

[37] Forbes MG, Forbes JF, Guay M. Regulatory control design for stochastic processes: shaping the probability density function. In Proc of the ACC 2003; 3998-4003.

[38] Forbes MG, Forbes JF, Guay M. Regulating discrete-time stochastic systems: focusing on the probability density function. Dynam Cont, Discrete Impuls Syst-Series B. 2004; 11(1): 81-100.

[39] Forbes MG, Guay M, Forbes JF. Control design for first-order processes: shaping the probability density of the process state. J Process Control. 2004; 14(4): 399-410.

[40] Francois G, Srinivasan B, Bonvin D, Hernandez J, Hunkeler H. Run-to-run adaptation of a semi-adiabatic policy for the optimization of an industrial batch polymerization process. I&EC Res. 2004; 43(23): 7238-42.

[41] Frank PM, Ding SX. Survey of robust residual generation and evaluation methods in observer-based fault detection systems. J Process Control. 1997; 7(6): 403-24.

[42] Gadkar K, Varner J, Doyle III FJ. Model identification of signal transduction networks from data using a state regulator problem. IEE Proc Syst Biol. 2005; 2(1): 17-30.

[43] Gahinet P, Nemirovski A, Laub AJ, *et al.* LMI control toolbox. The Math Works Inc, 1995.

[44] Garcia CE, Morari M. Internal model control. 2. Design procedure for multivariable systems. Ind Eng Chem Process Des Dev. 1985; 24(2): 472-84.

[45] Garcia CE, Morari M. Internal Model Control. 3. Multivariable control law computation and tuning guidelines. Ind Eng Chem Process Des Dev. 1985; 24(2): 484-94.

[46] Gattu G, Zafiriou E. Nonlinear quadratic dynamic matrix control with state estimation. Ind Eng Chem Res. 1992; 31(4): 1096-104.

[47] Ge M, Chiu MS, Wang QG. Robust PID control design via LMI approach. J Process Control. 2002; 12(1): 3-13.

[48] Gertler JJ. Fault detection and diagnosis in engineering systems. New York: Marcel Dekker, 1998.

[49] Ge SS, Hang CC, Lee TH, Zhang T. Stable adaptive neural network control. Boston, MA, Kluwer Academic, 2001.

[50] Ge SS, Lee TH, Harris CJ. Adaptive neural network control of robotic manipulators. Singapore, World Scientific, 1998.

[51] Ge SS, Lee TH, Li GY, Zhang J. Adaptive NN control for a class of discrete-time nonlinear systems. Int J Control. 2003; 76(4): 334-54.

[52] Gommeren HJC, Heitzmann DA, Moolenaar JAC, Scarlett B. Modelling and control of a jet mill plant. Powder Technol. 2000; 108(2): 147-54.

[53] Goodwin GC, Sin KS. Adaptive filtering: prediction and control. Englewood Cliffs, NJ,Prentice-Hall, 1984.

[54] Gregorcic G, Lightbody G. Local model network identification with Gaussian processes. IEEE Trans Neural Netw. 2007; 18(5): 1404-23.

[55] Grigoriadis KM, Watson JT. Reduced-order $H_\infty$ and $L_2/L_\infty$ filtering via linear matrix inequalities. IEEE Trans Aerosp Electron Syst. 1997; 33(4): 1326-38.

[56] Guo L. $H_\infty$ output feedback control for delay systems with nonlinear and parametric uncertainties. IEE Proc Control Theory Applic. 2002; 149(3): 226-36.

[57] Guo L. Guaranteed cost control of uncertain discrete-time delay systems using dynamic output feedback. Trans Inst Measure Control. 2002; 24(5): 417-30.

[58] Guo L. Statistic tracking control: a multi-objective optimization algorithm. In Proc 3rd International Symposium on Neural Networks. 2006; 962-7.

[59] Guo L, Chen WH. Output feedback $H_\infty$ control for a class of uncertain nonlinear discrete-time delay systems. Trans Inst Measure Control. 2003; 25(2): 107-21.

[60] Guo L, Malabre M. Robust $H_\infty$ control for descriptor systems with non-linear uncertainties. Int J Control. 2003; 76(12): 1254-62.

[61] Guo L, Wang H. Pseudo-PID tracking control for a class of output PDFs of general non-Gaussian stochastic systems. In Proc of the American Control Conference. 2003; 362-7.

[62] Guo L, Wang H. Applying constrained nonlinear generalized PI strategy to PDF tracking control through square root B-spline models. Int J Control. 2004; 77(17): 1481-92.

[63] Guo L, Wang H. PID controller design for output PDFs of stochastic systems using linear matrix inequalities. IEEE Trans Syst, Man Cybern B 2005; 35(1): 65-71.

[64] Guo L, Wang H. Generalized discrete-time PI control of output PDFs using square root B-spline expansion. Automatica. 2005; 41: 159-62.

[65] Guo L, Wang H. Minimum entropy filtering for multivariate stochastic systems with non-Gaussian noises. In Proc of the ACC. 2005; 315-20.

[66] Guo L, Wang H. Fault detection and diagnosis for general stochastic systems using B-spline expansions and nonlinear filters. IEEE Trans Circuits Syst I. 2005; 52(8): 1644-52.

[67] Guo L, Wang H. Minimum entropy filtering for multivariate stochastic systems with non-Gaussian noises. IEEE Trans Automatic Control. 2006; 51(4): 695-700.
[68] Guo L, Wang H. Entropy optimization filtering for fault isolation of non-Gaussian systems. In Proc of 6th IFAC symposium on fault detection, supervision and safety of technical processes. 2006; 432-7.
[69] Guo L, Wang H, Wang AP. Optimal probability density function control for NARMAX stochastic systems. Automatica. 2008; 44: 1904-11.
[70] Guo L, Yang F, Fang J. Multiobjective filtering for nonlinear time-delay systems with nonzero initial conditions based on convex optimizations. Circuit Syst Sign Process. 2006; 25(5): 591-607.
[71] Guo L, Zhang YM, Feng CB. Generalized $H_\infty$ performance and mixed $H_2/H_\infty$ optimization for time delay systems. In Proc of the International Conference on Automation, Robotics and Computer Vision. 2004; 36-41.
[72] Guo L, Zhang YM, Wang H. Fault diagnostic filtering using stochastic distributions in nonlinear generalized $H_\infty$ setting. In Proc of 6th IFAC symposium on fault detection, supervision and safety of technical processes. 2006; 216-21.
[73] Guo L, Zhang YM, Wang H, Fang JC. Observer based optimal fault detection and diagnosis using conditional probability distributions. IEEE Trans Signal Process. 2006; 54(10): 3712-19.
[74] Hayakawa T, Haddad WM, Hovakimyan N. Neural network adatpive control for a class of nonlinear dynamical systems with asymptotic stability guarantees. IEEE Trans Neural Netw. 2008; 19(1): 80-9.
[75] Haykin S. Neural networks: a comprehensive foundation. NewYork, IEEE Press, 1994.
[76] Hibey JL, Charalambous CD. Conditional densities for continuous time nonlinear hybrid systems with applications to fault detection. IEEE Trans Automatic Control. 2003; 44(11): 2164-70.
[77] Hunt KJ, Sbarbaro D, Zbikowski R, Gawthrop PJ. Neural networks for control systems: a survey. Automatica. 1992; 28(6): 1083-1112.
[78] Hounslow MJ, Ryall RL, Marshall VR. A discretized population balance for nucleation, growth and aggregation. AIChE J. 2004; 34(11): 1821-32.
[79] Huang SN, Tan KK, Lee TH. A combined PID adaptive controller for a class of nonlinear systems. Automatica. 2001, 37(4): 611-8.
[80] Immanuel CD, Cordeiro CF, Sundaram SS, Doyle III FJ. Population balance PSD model for emulsion polymerization with steric stabilizers. AIChE J. 2003; 49(6): 1392-404.
[81] Immanuel CD, Cordeiro CF, Sundaram SS, Meadows ES, Crowley TJ, Doyle III FJ. Modeling of particle size distribution in emulsion co-polymerization: comparison with experimental data and parametric sensitivity studies. Comput Chem Eng. 2002; 26(7): 1133-52.
[82] Immanuel CD, Doyle III FJ. Computationally-efficient solution of population balance models incorporating nucleation, growth and coagulation. Chem Eng Sci. 2003; 58(16): 3681-98.

[83] Immanuel CD, Doyle III FJ. Solution technique for a multi-dimensional population balance model describing granulation processes. Powder Technol. 2005; 156(2): 213-25.
[84] Isermann R, Balle P. Trends in the application of model based fault detection and diagnosis of technical process. In Proc 13th International Federation of Automatic Control 1996; 1-12.
[85] Jiang B, Chowdhury FN. Fault estimation and accommodation for linear MIMO discrete time systems. IEEE Trans Control Syst Technol. 2005; 13(3): 493-9.
[86] Jiang B, Wang JL, Soh YC. An adaptive technique for robust diagnosis of faults with independent effect on system outputs. Int J Control. 2002; 75(11): 792-802.
[87] Jones DG, Wang H. A new nonlinear optimal control strategy for paper formation. J Measure Control. 1999; 32(8): 241-5.
[88] Kabore P, Baki H, Yue H, Wang H. Linearized controller design for the output probability density functions of non-Gaussian stochastic systems. Int J Automation Comput. 2005; 2(1): 67-74.
[89] Kalani A, Christofides PD. Nonlinear control of spatially inhomogeneous aerosol processes. Chem Eng Sci. 1999; 54(3): 2669-78.
[90] Kalani A, Christofides PD. Simulation, estimation and control of size distribution in aerosol processes with simultaneous reaction, nucleation, condensation and coagulation. Comput Chem Eng. 2002; 26: 1153-69.
[91] Karray F, Gueaieb W, Sharhan SA. The hierarchical expert tuning of PID controllers using tools of soft computing. IEEE Trans Syst, Man Cybernetics B. 2002; 32(1): 77-90.
[92] Karny M. Towards fully probabilistic control design. Automatica. 1996; 32(2): 1719-22.
[93] Karny M, Bohm J, Guy TV, Nedoma P. Mixture-based adaptive probabilistic control. Int J Adaptive Control Signal Process. 2003; 17(2): 119-32.
[94] Kim E, Lee H. New approaches to relaxed quadratic stability condition of fuzzy control systems. IEEE Trans Fuzzy Syst. 2000; 8(5): 523-34.
[95] Kiparissides C, Seferlis P, Mourikas G, Morris AJ. Online optimization control of molecular weight properties in batch free-radical polymerisation reactors. Ind Eng Chem Res 2002; 41(24): 6120-31.
[96] Kosmatopoulos EB, Polycarpou MM, Christodoulou MA, Ioannou PA. High-order neural network structures for identification of dynamical systems. IEEE Trans Neural Netw. 1995; 6(2): 422-31.
[97] Kozub DJ, MacGregor JF. Feedback control of polymer quality in semi-batch polymerization reactors. Chem Eng Sci. 1992; 47(4): 929-42.
[98] Kumar S, Ramkrishna D. On the solution of population balance equations by discretization I. A fixed pivot technique. Chem Eng Sci. 1996; 51(8): 1311-32.
[99] Kumar S, Ramkrishna D. On the solution of population balance equations by discretization II. A moving pivot technique. Chem Eng Sci. 1996; 51(8): 1333-44.

[100] Lee K, Lee JH, Yang DR, Mahoney AW. Integrated run-to-run and on-line model-based control of particle size distribution for a semi-batch precipitation reactor. Comput Chem Eng. 2002; 26(7): 1117-31.
[101] Lee KR, Kim JH, Jeung ET, Park HB. Output feedback robust $H_\infty$ control of uncertain fuzzy dynamic systems with time-varying delay. IEEE Trans Fuzzy Syst. 2000; 8(12): 657-64.
[102] Lee KS, Lee JH. Iterative learning control-based batch process control technique for integrated control of end product properties and transient profiles of process variables. J Process Control. 2003; 13(7): 607-21.
[103] Lewis FL, Yesildirek A, Liu K. Multilayer neural network robot controller with guaranteed tracking performance. IEEE Trans Neural Netw. 1996; 7(2): 388-98.
[104] Lin CM, Chen LY, Chen CH. RCMAC hybrid control for MIMO uncertain nonlinear systems using sliding-mode technology. IEEE Trans Neural Netw. 2007; 18(3): 708-20.
[105] Lin CM, Hsu CF. Recurrent neural network based adaptive backstepping control for induction servomotors. IEEE Trans Ind Electron. 2005; 52(6): 1677-84.
[106] Lin C, Wang QG, Lee TH. An improvement on multivariable PID controller design via iterative LMI approach. Automatica. 2004; 40(3): 519-25.
[107] Lin C, Wang QG, Lee TH. $H_\infty$ output tracking control for nonlinear systems via T-S fuzzy model approach. IEEE Trans Syst, Man Cybern B. 2006; 36(2): 450-7.
[108] Litster JD, Smith DJ, Hounslow MJ. Adjustable discretized population balance for growth and aggregation. AIChE J. 1995; 41(3): 591-603.
[109] Li T, Yi Y, Guo L, Wang H. Delay-dependent fault detection and diagnosis using B-spline neural networks and nonlinear filters for time-delay stochastic systems. Neural Comput Applic. 2008; 17(4): 405-11.
[110] Liu J, Wang JL, Yang GH. Reliable guaranteed variance filtering against sensor failures. IEEE Trans Signal Process. 2003; 51(5): 1403-11.
[111] Li X, Souza CE. Delay-dependent robust stability and stabilization of uncertain linear delay systems: a linear matrix inequality approach. IEEE Trans Automatic Control. 1997; 42(8): 1144-48.
[112] Mahoney AW, Ramkrishna D. Efficient solution of population balance equations with discontinuities by finite elements. Chem Eng Sci. 2002; 57(7): 1107-19.
[113] Ma J, Feng G. An approach to $H_\infty$ control of fuzzy dynamic systems. Fuzzy Sets Syst. 2003; 137(3): 367-86.
[114] Marino R, Tomei P. Nonlinear control design: geometric, adaptive and robust. Englewood Cliffs, NJ, Prentice Hall, 1995.
[115] Masubuchi I, Kamitane Y, Ohara A, Suda N. $H_\infty$ control for descriptor system: a matrix inequalities approach. Automatica. 1997; 33(4): 669-73.
[116] Mattei M. Robust multivariable PID controllers for linear parameter varying systems. Automatica. 2001, 37(12), 1997-2003.

[117] Ma Z, Merkus HG, de Smet JGAE, Heffels C, Scarlett B. New developments in particle characterization by laser diffraction: size and shape. Powder Technol. 2000; 111(1): 66-78.

[118] Narendra KS, Parthasarathy K. Identification and control of dynamical systems using neural networks. IEEE Trans Neural Netw. 1990; 1(1): 4-27.

[119] Owens C, Zisser H, Jovanovic L, Srinivasan B, Bonvin D, Doyle III FJ. Run-to-run control of blood glucose concentrations for people with type 1 diabetes mellitus. IEEE Trans Biomed Eng. 2006; 53(6): 996-1005.

[120] Park MY, Dokucu MT, Doyle III FJ. Regulation of the emulsion particle size distribution using partial least squares model-based predictive control. Ind Eng Chem Res. 2004; 43(23): 7227-37.

[121] Patton RJ, Chen J. Control and dynamic systems: robust fault detection and isolation (FDI) systems. London, Academic, 1996.

[122] Polycarpou MM, Mears MJ. Stable adaptive tracking of uncertain systems using nonlinearly parameterized on-line approximators. Int J Control. 1998; 70(3): 363-84.

[123] Poznyak AS, Lennart L. On-line identification and adaptive trajectory tracking for nonlinear stochastic continuous time systems using differential neural networks. Automatica. 2001; 37(8): 1257-68.

[124] Poznyak AS, Yu W, Sanchez EN, Perez JP. Nonlinear adaptive trajectory tracking using dynamic neural networks. IEEE Trans Neural Netw. 1999; 10(6): 1402-11.

[125] Ramkrishna D. Population balances. San Diego, Academic Press, 2000.

[126] Ren XM, Rad AB. Identification of nonlinear systems with unknown time delay based on time-delay neural networks. IEEE Trans Neural Netw. 2007; 18(5): 1536-41.

[127] Ren XM, Rad AB, Chan PT, Lo WL. Identification and control of continuous-time nonlinear systems via dynamic neural networks. IEEE Trans Ind Electron. 2003; 50(3): 478-86.

[128] Ribeiro A, Giannakis GB. Bandwidth constrained distributed estimation for wireless sensor networks-Part II: unknown probability density function. IEEE Trans Signal Process. 2006; 54(7): 2784-96.

[129] Ricker NL. Model predictive control with state estimation. Ind Eng Chem Res. 1990; 29(3): 374-82.

[130] Rovithakis A. Robust neural adaptive stabilization of unknown systems with measurement noise. IEEE Trans Syst, Man Cybern B. 1999; 29(3): 453-9.

[131] Rubio JJ, Yu W. Stability analysis of nonlinear system identification via delayed neural networks. IEEE Trans Circuits Syst II. 2007; 54(2): 161-5.

[132] Saldivar E, Ray WH. Control of semicontinuous emulsion copolymerization reactors. AIChE J. 1997; 43(8): 2021-33.

[133] Scherer C, Weiland S. Lecture notes DISC course on linear matrix inequalities in control. Delft, The Netherlands, Dutch Institute of Systems and Control, 2000.

[134] Scott I, Mulgrew B. Nonlinear system identification and prediction using orthonormal functions. IEEE Trans Signal Process. 1997; 45(7): 1842-53.

[135] Semino D, Ray WH. Control of systems described by population balance equations-II. emulsion polymerization with constrained control action. Chem Eng Sci. 1995; 50(11): 1825-39.
[136] Seo YB, Choi JW. Stochastic eigenvalues for LTI systems with stochastic modes. In Proc of the 40th SICE Annual Conference. 2001; 142-5.
[137] Shalom YB, Li XR, Kirubarajan T. Estimation with applications to tracking and navigation. New York, John Wiley and Sons, 2001.
[138] Shalom YB, Li XR. Nonlinear filter design using Fokker-Planck-Kolmogorov probability density evolutions. IEEE Trans Aerosp Electron Syst. 2000; 36(1): 309-15.
[139] Smook GA. Handbook for pulp and paper technologists. Vancouver, Angus Wilde Publications Inc., 1992.
[140] Soares JBP, Kim JD, Rempel GL. Analysis and control of the molecular weight and chemical composition distributions of polyolefins made with metallocene and Ziegler-Natta catalysts. Ind Eng Chem Res. 1997; 36(4): 1144-50.
[141] Socha L, Blachuta M. Application of linearization methods with probability densisty criteria in control problems. In Proc of the American Control Conference. 2000: 2775-79.
[142] Srinivasan B, Primus CJ, Bonvin D, Ricker NL. Run-to-run optimization via control of generalized constraints. Control Eng Practice. 2001; 9(8): 911-9.
[143] Stelling J, Sauer U, Szallasi Z, Doyle III FJ, Doyle J. Robustness of cellular functions. Cell. 2004; 118(6): 675-85.
[144] Stoorvogel AA, Niemann HH, Saberi A, Sannuti P. Optimal fault signal estimation. Int J Robust Nonlinear Control. 2002; 12(8): 697-727.
[145] Storti G, Carra S, Morbidelli M, Vita G. Kinetics of multimonomer emulsion polymerization, the pseudo-homopolymerization approach. J. Appl Polym Sci. 1989; 37(9): 2443-67.
[146] Sun X, Yue H, Wang H. Modelling and control of the flame temperature distribution using probability density function shaping. Trans Inst Measure Control. 2006; 28(5): 401-28.
[147] Tanaka K, Ikeda T, Wang HO. Robust stabilization of a class of uncertain nonlinear systems via fuzzy control: quadratic stabilizability, $H_\infty$ control theory and linear matrix inequalities. IEEE Trans Fuzzy Syst. 1996; 4(1): 1-13.
[148] Taniguchi T, Tanaka K, Wang HO. Fuzzy descriptor systems and nonlinear model following control. IEEE Trans Fuzzy Syst. 2000; 8(4): 442-52.
[149] Takagi T, Sugeno M. Fuzzy identification of systems and its applications to modeling and control. IEEE Trans Syst, Man Cybern. 1985; 15(1): 116-32.
[150] Tseng CS, Chen BS. Multiobjective PID control design in uncertain robotic systems using neural network elimination scheme. IEEE Trans Syst, Man Cybern A. 2001, 31(6), 632-44.
[151] Tseng CS, Chen BS, Uang HJ. Fuzzy tracking control design for nonlinear dynamic systems via T-S fuzzy model. IEEE Trans Fuzzy Syst. 2001, 9(3): 381-92.

[152] Utkin VI. Sliding modes in control and optimization. New York, Springer-Verlag, 1992.
[153] Vicente M, BenAmor S, Gugliotta LM, Leiza JR, Asua JM. Control of molecular weight distribution in emulsion polymerization using on-line reaction calorimetry. Ind Eng Chem Res. 2001; 40(1): 218-27.
[154] Vicente M, Sayer C, Leiza JR, Arzamendi G, Lima EL, Pinto JC, Asua JM. Dynamic optimization of non-linear emulsion copolymerization systems open-loop control of composition and molecular weight distribution. Chem Eng J. 2002; 85(2): 339-49.
[155] Wang AP, Wang H. Minimum entropy control using B-spline square root models. IEE Control Theory Applic D. 2004; 151(4): 422-8.
[156] Wang H, Baki H, Kabore P. Control of bounded stochastic distributions using square root models: an applicability study. Trans Inst Measure Control. 2001; 23(1): 51-68.
[157] Wang H. Bounded dynamic stochastic distributions modeling and control. Springer-Verlag, 2000.
[158] Wang H. Control of Conditional output PDF for general nonlinear and non-Gaussian stochastic systems. IEE Control Theory Applic D. 2003; 150(1): 55-60.
[159] Wang H. Iterative learning based B-splines for output probability density function control. Keynote presentation at international conference on applied cybernetics, 2005.
[160] Wang H, Kabore P, Baki H. Lyapunov based design for bounded dynamic stochastic distribution control. IEE Control Theory Applic D. 2001; 148(3): 245-50.
[161] Wang H, Lin W. Applying observer based FDI techniques to detect faults in dynamic and bounded stochastic distributions. Int J Control. 2000; 73(15): 1424-36.
[162] Wang H. Minimum entropy control for non-Gaussian dynamic stochastic systems. IEEE Trans Automatic Control. 2002; 47(2): 398-403.
[163] Wang H. Model reference adaptive control of the output probability density functions for unknown linear dynamic stochastic systems. Int J Syst Sci. 1999; 30: 707-15.
[164] Wang H. Robust control of the output probability density functions for multivariable stochastic systems with guaranteed stability. IEEE Trans Automatic Control. 1999; 44(11): 2103-07.
[165] Wang H, Wang AP, Wang Y. An online estimation algorithm for the unknown probability density functions of random parameters in stochastic ARMAX systems. IEE Control Theory Applic D. 2006; 153: 462-8.
[166] Wang H, Yue H. A rational spline model approximation and control of output probability density functions for dynamic stochastic systems. Trans Inst Measure Control. 2003; 25(2): 93-105.
[167] Wang H, Yue H. Output PDF control of stochastic distribution systems: modelling, control and application. Control Eng China. 2003; 10(3): 193-7.

[168] Wang H, Zhang JH. Bounded stochastic distribution control for pseudo ARMAX systems. IEEE Trans Automatic Control. 2001; 46(3): 486-90.
[169] Wang H, Zhang JF, Yue H. Multi-step predictive control of a PDF-shaping problem. Automatica Sinica. 2005; 31(2): 274-9.
[170] Wang LX. Adaptive fuzzy systems and control design and stability analysis. New Jersey, Prentice-Hall, 1994.
[171] Wang LX. Stable adaptive fuzzy control of nonlinear systems. IEEE Trans Fuzzy Syst. 1993; 1(2): 146-55.
[172] Wang Y, Wang H. Nonlinear one-step-ahead predictive mean control of bounded dynamic stochastic systems with guaranteed stability. Int J Syst Sci. 2004; 35: 97-108, .
[173] Wang Y, Wang H. Output PDFs control of linear stochastic systems with arbitrarily bounded random parameters, a new application of the Laplace transforms. In Proc of the American Control Conference. 2002; 4262-67.
[174] Wang ZD, Daniel WCH, Liu XH. A note on the robust stability of uncertain stochastic fuzzy systems with time-delays. IEEE Trans Syst, Man Cyberne A. 2004; 34(4): 570-6.
[175] Wang ZD, Liu YR, Li MZ, Liu XH. Stability analysis for stochastic cohen-grossberg neural networks with mixed time delays. IEEE Trans Neural Netw. 2006; 17(3): 814-20.
[176] Wang ZD, Yang FW. Robust filtering for uncertain linear systems with delayed states and outputs. IEEE Trans Circuits Syst I. 2002; 49(1): 125-30.
[177] Ying H. Analytical analysis and feedback linearization tracking control of the general Takagi-Sugeno fuzzy dynamic systems. IEEE Trans Syst, Man Cybern C. 1999; 29(2): 290-8.
[178] Xiong ZH, Zhang J. A batch-to-batch iterative optimal control strategy based on recurrent neural network models. J Process Control. 2005; 15(1): 11-21.
[179] Xu SY, Lam J, Ho WC. A new LMI condition for delay-dependent asymptotic stability of delayed hopfield neural networks. IEEE Trans Circuits Syst II. 2006; 53(3): 230-4.
[180] Xu SY, Lam J, Mao XR. Delay-dependent $H_\infty$ control and filtering for uncertain markovian jump systems with time-varying delays. IEEE Trans Circuits Syst I. 2007; 54(9): 2070-7.
[181] Yi Y, Guo L, Wang H. Adaptive statistic tracking control based on two steps neural networks with time delays. IEEE Trans Neural Netw. 2009; 20(3):420-9.
[182] Yi Y, Li T, Guo L, Wang H. Adaptive tracking control for the output PDFs based on dynamic neural networks. In Proc of the International Symposium on Neural Networks. 2007: 93-101.
[183] Yi Y, Li T, Guo L, Wang H. Statistic tracking strategy for non-Gaussian systems based on PID controller structure and LMI approach. Dynam Continuous, Discrete Impulsive Syst B. 2008; 15: 859-72.
[184] Yi Y, Shen H, Guo L. Statistic PID tracking control for non-Gaussian stochastic systems based on T-S fuzzy model. Int J Automation Comput. 2009; 6(1): 81-7.

[185] Yoon BJ, Vaidyanathan PP. A multirate DSP model for estimation of discrete probability density functions. IEEE Trans Signal Process. 2005; 53(1): 252-64.
[186] Yue H, Jiao J, Brown EL, Wang H. Real-time entropy control of stochastic systems for an improved paper web formation. J Measure Control. 2001; 34(5): 134-9.
[187] Yue H, Wang H. Minimum entropy control of closed loop tracking errors for dynamic stochastic systems. IEEE Trans Automatic Control. 2003; 48(1): 118-22.
[188] Yue H, Wang H. Recent developments in stochastic distribution control, a review. J Measure Control. 2003; 36: 209-15. 2003.
[189] Yu W, Li XO. Some new results on system identification with dynamic neural networks. IEEE Transn Neural Netw. 2001; 12(2): 412-17.
[190] Yu W. Passivity analysis for dynamic multilayer neuro identifier. IEEE Trans Circuits Syst I. 2003; 50(1): 173-8.
[191] Zhang HG, Wang ZS, Liu DR. Global asymptotic stability of recurrent neural networks with multiple time-varying delays. IEEE Trans Neural Netw. 2008; 19(5): 855-73.
[192] Zhang HG, Yang J, Su CY. T-S fuzzy model based robust $H_\infty$ design for networked control systems with uncertainties. IEEE Trans Ind Informatics. 2007; 3(4): 289-301.
[193] Zhang HG, Yang DD, Chai TY. Guaranteed cost networked control for T-S fuzzy systems with time delays. IEEE Trans Syst, Man Cybern C. 2007, 37(2): 160-72.
[194] Zhang TP, Ge SS. Adaptive neural control of MIMO nonlinear state time-varying delay systems with unknown dead-zones and gain signs. Automatica. 2007; 43: 1021-33.
[195] Zhang X, Polycarpou M, Parisini T. A robust detection and isolation scheme for abrupt and incipient faults in nonlinear systems. IEEE Trans Automatic Control. 2002; 47(7): 576-93.
[196] Zhang YM, Yu HS, Guo L. Using guaranteed cost filters for fault detection of discrete-time stochastic distribution systems with time delays. In Proc of 6th World Congress on Intelligent Control Automation. 2006: 5521-25.
[197] Zhao Q, Xu Z. Design of a novel knowledge-based fault detection and isolation scheme. IEEE Trans Syst, Man Cybern B. 2004; 34(2): 1089-95.
[198] Zheng F, Wang QG, Lee TH. On the design of multivariable PID controllers via LMI approach. Automatica. 2002; 38(3): 517-26.
[199] Zheng F, Wang QG, Lee TH. Output tracking control of MIMO fuzzy nonlinear systems using variable structure control approach. IEEE Trans Fuzzy Syst. 2002; 10(6): 686-97.
[200] Zhong MY, Ye H, Ding SY, *et al.* Observer-based fast rate fault detection for a class of multirate sampled-data systems. IEEE Trans Automatic Control. 2007; 52: 520-5.
[201] Zhou D, Ye Y. Modern fault diagnosis and fault tolerant control (in Chinese). Beijing, China, Tsinghua University Press, 2000.

[202] Zhou K, Doyle JC. Essentials of robust control. Prentice-Hall, 2000.
[203] Zhou K, Doyle JC, Glover K. Essential of robust control. Englewood Cliffs, NJ, Prentice-Hall, 1997.

# Index

**Other titles published in this series (continued):**

*Soft Sensors for Monitoring and Control of Industrial Processes*
Luigi Fortuna, Salvatore Graziani, Alessandro Rizzo and Maria G. Xibilia

*Adaptive Voltage Control in Power Systems*
Giuseppe Fusco and Mario Russo

*Advanced Control of Industrial Processes*
Piotr Tatjewski

*Process Control Performance Assessment*
Andrzej W. Ordys, Damien Uduehi and Michael A. Johnson (Eds.)

*Modelling and Analysis of Hybrid Supervisory Systems*
Emilia Villani, Paulo E. Miyagi and Robert Valette

*Process Control*
Jie Bao and Peter L. Lee

*Distributed Embedded Control Systems*
Matjaž Colnarič, Domen Verber and Wolfgang A. Halang

*Precision Motion Control* (2nd Ed.)
Tan Kok Kiong, Lee Tong Heng and Huang Sunan

*Optimal Control of Wind Energy Systems*
Iulian Munteanu, Antoneta Iuliana Bratcu, Nicolaos-Antonio Cutululis and Emil Ceangă

*Identification of Continuous-time Models from Sampled Data*
Hugues Garnier and Liuping Wang (Eds.)

*Model-based Process Supervision*
Arun K. Samantaray and Belkacem Bouamama

*Diagnosis of Process Nonlinearities and Valve Stiction*
M.A.A. Shoukat Choudhury, Sirish L. Shah, and Nina F. Thornhill

*Magnetic Control of Tokamak Plasmas*
Marco Ariola and Alfredo Pironti

*Real-time Iterative Learning Control*
Jian-Xin Xu, Sanjib K. Panda and Tong H. Lee

*Deadlock Resolution in Automated Manufacturing Systems*
ZhiWu Li and MengChu Zhou

*Model Predictive Control Design and Implementation Using MATLAB®*
Liuping Wang

*Fault-tolerant Flight Control and Guidance Systems*
Guillaume Ducard

*Predictive Functional Control*
Jacques Richalet and Donal O'Donovan

*Fault-tolerant Control Systems*
Hassan Noura, Didier Theilliol, Jean-Christophe Ponsart and Abbas Chamseddine

*Control of Ships and Underwater Vehicles*
Khac Duc Do and Jie Pan

*Detection and Diagnosis of Stiction in Control Loops*
Mohieddine Jelali and Biao Huang

*Dry Clutch Control for Automotive Applications*
Pietro J. Dolcini, Carlos Canudas-de-Wit, and Hubert Béchart

*Nonlinear and Adaptive Control Design for Induction Motors*
Riccardo Marino, Patrizio Tomei and Cristiano M. Verrelli
Publication due July 2010

*Advanced Control and Supervision of Mineral Processing Plants*
Daniel Sbárbaro and René del Villar (Eds.)
Publication due October 2010

*Active Braking Control Design for Road Vehicles*
Sergio M. Savaresi and Mara Tanelli
Publication due October 2010